Copyright©2022

W0019570

ACKNOWLEDGEMENT

"Everyone around a Ph.D. student contributes to the completion of the dissertation ranging from the enthusiastic guidance of the supervisor, the encouragement of the close friend, to the unending support of the one's family".

The work presented in the thesis has been completed due to various contributions, help and support in many ways. Here is a great opportunity for me to acknowledge all those people who were directly or indirectly related to my thesis work. My initial gratitude goes to **Vivekananda Global University, Jaipur** for giving me this opportunity to carry out my research work. I would like to express my profound regards and sincere thanks to my respected supervisor **Dr. Ruchi Sharma**. Without her valuable guidance and support it was not possible for me to complete this thesis work. She is very simple and hardworking and believes in self-working concept whether the work is big or small. She provided me the right direction for the research and always supported in the difficult situations. Her guidance will be a lifelong lesson for me.

I am thankful to **Dr. K. Ram Bagaria,** Chief Patron, **Dr. K.R. Bagaria,** Vice-Chairperson, **Prof. (Dr.) M. Raisinghani,** Ex. Vice-Chairperson, **Er. Gaurav Bagaria,** Director, **Er. Onkar Bagaria,** CEO, **Prof. (Dr.) Vijay Vir Singh,** President, **Prof. (Dr.) Praveen Choudhry,** Registrar, **Prof. (Dr.) Baldev Singh** Dean Engineering, **Dr. Subodh Srivastava,** Dean In-Charge R&D, VGU, for providing all necessary support and facilities in University. I am thankful to **Mr. Rahul Sharma** for providing wonderful support during complete tenure of Ph.D.

I cannot forget role of **Mr. Mahaveer Kushwah**, faculty member PACE Academy, Kota for his contribution in building my foundation in Computer programming.

I am sincerely thankful to **Dr. Arun Kumar,** Associate Professor, JECRC University, Jaipur and **Er. Mohit Sharma** PhD scholar, MNIT, Jaipur for valuable support, guidance and inspiration. I also remember the encouragement from my

iv

colleague and friend **Mr. Rajendra Sharma**, Sr. IT Manager, Suresh Gyan Vihar University, Jaipur. They provided me all the possible help, guidance and suggestions and experimental support. I always felt free and comfortable to discuss my problems to them and got the right solutions.

I am also thankful to **Prof. Sanjiv Kumar**, former HOD, ECE, **Prof. (Dr.) D.P. Darmora**, former Principal, Vivekananda Institute of Technology, Jaipur and **Prof. (Dr.) Y.C. Sharma**, Former Dean R&D, VGU Jaipur for their guidance and motivation during work.

I want to acknowledge my respected parents, **Shri Ramesh Chand Sharma** (Retd. Sr. Teacher) and **Smt. Sushila Devi Sharma** for their encouragement, patience and support throughout my academic career and this long Ph.D. tenure. I am thankful to my loving brothers **Er. Neeraj Jaiman,** Manager, Production Planning and Control, D'Decore Exports Pvt. Ltd., Tarapur, Maharashtra and **Er. Vikas Jaiman**, Asst. Professor, Compucom Institute of Technology and Management, Jaipur for their support so that, I could put all my efforts, capability and time to perform research work.

At last, I am really grateful to my lovely wife **Mrs. Devki Sharma** for her continuous help, support and encouragement to complete the thesis work at final stage. I acknowledge my son **Saket Jaiman** whose cute smile and attractive activities always provide me a freshness and happy mood.

Finally, I sincerely thank to **VGU and VIT family** for supporting me in every possible way.

Akash Jaiman

CONTENTS

LIST OF FIGURES

LIST OF TABLES

LIST OF ABBREVIATIONS

IOT	:	Internet of Things
AI	:	Artificial Intelligence
ML	:	Machine Learning
CNN	:	Convolutional Neural Network
ANN	:	Artificial Neural Network
DL	:	Deep Learning
SVM	:	Support Vector Machine
WSA	:	Wireless Sensor Network
RF	:	Radio Frequency
PAMS	:	Precision Agricultural Management System
AMIS	:	Agricultural Management Information System
AIT	:	Agricultural Information Technology
GWL	:	Ground Water Level
SWL	:	Surface Water Level
UAV	:	Unmanned Arial Vehicle
TDR	:	Time Domain Reflectometry
LRF	:	Laser Range Fielder
GIS	:	Geographic Information System
JSON	:	JAVA Script Object Notation

ABSTRACT

The aim of smart farming is to make every feature of farming more consistent, foreseeable, and supportable. The role of the different agri-technologies is to close this information gap by providing accurate measurements of the aspects that regulate farming consequences. Smart Farming is characterized by the application of contemporary information and communication technologies (ICT) into farming, foremost to what can be termed a Third Green Revolution. Following the mutinies in plant breeding and genetics, the Third Green Revolution is enthralling the farming domain through the mutual solicitation of ICT resolutions such as precision apparatus, the Internet of Things (IoT), sensors, geo-positioning systems, Big Data, and robotics. Artificial Intelligence (AI) based technologies help to improve efficiency in all the fields and also manage the challenges faced by various industries, including the various fields in the agricultural sector like crop yield, irrigation, soil content sensing, crop monitoring, weeding, and crop establishment. The proposed systems help to improve the overall harvest quality and accuracy – known as precision agriculture. AI technology helps in detecting disease in plants, pests and poor nutrition in farms. AI sensors can detect and target weeds and then decide which herbicide to apply within the region. The system will traverse the fields and work autonomously to respond to the needs of crops, and perform weeding, watering, pruning, and harvesting functions guided by their own collection of sensors, navigation, and crop data. Furthermore, this work surveys the work of many researchers to get a brief overview of the current implementation of automation in agriculture, specifically the weeding systems through robots and drones. The various soil water sensing methods are discussed, along with two automated weeding techniques. The implementation of drones is being discussed. The various methods used by drones for spraying and crop-monitoring are also discussed in this work. The main concern of this work is to audit the various applications of artificial intelligence in agriculture, such as irrigation, weeding, and spraying, with the help of sensors and other means embedded in robots and drones. These technologies save the excess use of water, pesticides, and herbicides, maintain the fertility of the soil, help with the efficient use of manpower, and improve the quality. Additionally, screening tests are performed to detect potential health disorders or diseases in people who do not have any symptoms of disease. The goal is early detection and lifestyle changes or surveillance, to reduce the risk of disease, or to detect it early enough to treat it most effectively. Plant leaves can be used to effectively detect plant diseases. However, the number of images of unhealthy leaves collected from various plants is usually unbalanced. However, it is difficult to detect diseases using such an unbalanced dataset as that described in this work.

Chapter 1
INTRODUCTION TO SMART FARMING IN AGRICULTURE

1.1 Introduction

Nowadays, we are enclosed by a lot of "splendid" sensors and wise systems that are reliably between related through the Internet and cloud organizes; this is the Internet of Things (IoT) perspective that gives bleeding edge developments in the entire social and advantageous regions of the overall population [1]. Thinking about the general commercial center, associations battle to expand their profitability and financial framework by smoothing out costs, time, and resources and, all the while, attempting to improve the organization's character and the thing combination offered to customers. The thought towards adequacy and gainful upgrades is wanted similarly in the cultivating region, where the inventive components and the resource the board impact crop types, water frameworks, and disinfestations whole; Working on such creation rhythms with no modified control is likely beginning to bring resource waste, destroyed or abandoned yields, and dirtied and crushed grounds. Imaginative advances can be useful to tackle problems, for example the normal practicality, waste reduction and soil improvement. The get-together and the assessment of cultivating data, which join different and heterogeneous elements, are of critical premium for the possibility of making creation techniques, mindful of the organic framework and its resources (improvement of water framework and planting as indicated by soil history and infrequent cycles), the ID of convincing and non-powerful components, the possibility of doing feature examination similar to the guess of future hard-insightful information. The possibility of changing harvests under unequivocal conditions, and finally the ability to increase mechanical theory by limiting and envisioning dissatisfaction and machine replacements. The development of false reasoning supports a variety of zones to enhance productivity and efficiency. Human analysis plans help to overcome common challenges in every area. Similarly, AI in the agro-industry supports farmers to improve their profitability and reduce compromising biological impacts. The corporate culture has unequivocally and clearly recognized AI in its preparation to change the overall

1

outcome. The restored insight displaces the way our food is transported where the radiation from the farmland has decreased by 20%. Changing the AI progress makes it possible to control and manage any avoided base condition [2]. The use of artificial information in agriculture allows farmers to capture data encounters, such as weather, precipitation, wind speed and solar radiation [3]. The evaluation of critical features data, offers a predominant connection of ideal outcomes [4]. The most surprising part of the realization of AI in agro-industry is that it will not abstain from the places of human peasants but will enhance their cycles [5].

- AI is a more viable way to transport, harvest and sell core yields.
- The use of AI emphasizes controlling under-yields and improving the potential for healthy crop creation.
- Advances in AI have reinvigorated agribusiness associations to make them even more successful.
- AI is used in applications, for example, changing robotic machines for the intended environment and identifying diseases or bugs.
- Artificial knowledge can improve board practices, as needed, and help a variety of technology associations integrate assets into estimates that are growing in importance in agribusiness.
- AI play plans may address farmers' problems, such as climate assortment, an invasion of irritations and weeds that reduces yields.

The flow diagram for intelligent Agriculture using AI is shown in Fig.1.1

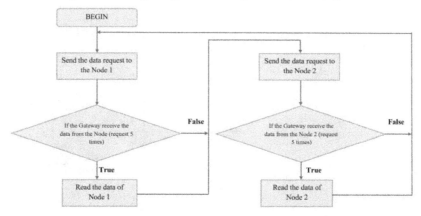

Figure 1.1: Flowchart of smart farming using AI

2

AI technology promptly fixes problems while recommending specific measures to overcome them. AI effectively tracks information in order to rapidly find solutions [6]. Let's understand how AI is being employed in farming to improve outcomes with a minimum environmental cost. Implementation of AI can identify an illness with 98% accuracy. As a result, AI helps farmers monitor fruits and vegetables by adjusting light to speed up production. Advanced artificial intelligence helps farmers stay up-to-date when it comes to weather predictions. Forecasted/ forecasted data helps farmers improve yields and profits without risking the harvest. Analysis of the data generated allows the farmer to take a precautionary approach by understanding and learning with artificial intelligence. The implementation of such a practice enables an informed decision to be taken on time [7]. The use of AI is an efficient way to identify or monitor potential deficiencies and nutrient deficiencies in the soil. With the image recognition approach, the AI identifies possible faults thanks to the images captured by the camera. With the help of the AI deep learning application are developed to analyze plant life patterns in agriculture. These AI-based applications provide insight into soil defects, plant pests and diseases. Farmers can use AI to manage weeds through the application of computer vision, robotics and machine learning. Using AI, data is collected to control weeds that help farmers spray chemicals [8]. It directly reduced the use of the chemical spraying a full field. As a result, AI reduces the use of herbicides in the field compared to the normal amount of sprayed chemicals [9]. The Smart Agriculture scenario is shown at Fig.1.2.

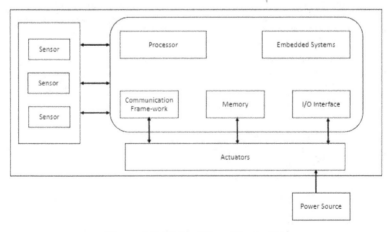

Figure 1.2 Basic of Smart Agriculture

3

Knowledge rebuilding has allowed agricultural robots to help farmers find more sustainable ways to protect their crops from weeds. It also contributes to overcoming the challenge of the job. PC-based knowledge robots in the crop field are able to harvest crops at a higher volume and faster than human experts [10]. Using PC vision helps to verify the weed and sprinkle it. As a result, AI helps farmers find more profitable ways to protect their crop from weeds. Today, the advances controlled by AI serve to regulate a few inspirations of adventures. Man-made insight is being utilized in the territories, for instance, cash, transport, clinical consideration, and now in agriculture. The foresight reproduced helps farmers to notice their yields without the necessity of invigilate before long in the property. Various new organizations and attempts anticipate a progression of AI within culture. The simulated overview reexamines the standard illustration of the culture [11]. The destiny of AI in culture is to move forward in the progressive change of commitment with a trailing edge approaches. Today, the majority of new agribusiness organizations are modifying their artificial intelligence approach to manage the increased profitability of agricultural creation. The market research report stated that the overall size of the artificial intelligence (AI) market in agriculture is needed to reach US$1,550 million before the completion of 2025. The achievement of philosophical AI could help distinguish disease or climate change earlier and respond strongly [12]. Agri-food associations, with the help of AI, are putting provincial data in place to reduce hostile results. With the general population impact, there is a speedy extension in the interest for food and green stocks and advancement connection to improve yield, cost-sufficiency, and nature of harvests/cultivating things being conveyed with new development, for instance, the Internet of Things (IoT) and Artificial Intelligence [13]. There is a need to assemble yield, reasonability and redesigned formation of land per unit district taken under idea. It is imperative to acknowledge new progressions to vanquish these issues. There are various points of interest related with the foundation of new advances which include: extended benefit, genuine yield assignment, Crop plan proposition, authentic use of resources, for instance, Fertilizers and waste products using the strategy for Automation and AI model. The arising idea of savvy cultivating that makes agribusiness more productive and powerful with the assistance of high-accuracy calculations. The component that drives it is Machine Learning — the logical field that enables machines to learn without being carefully customized [14]. It has arisen along with large information advances and elite registering to set out new open doors to disentangle, evaluate, and comprehend information concentrated cycles in horticultural operational conditions. AI is wherever all through the entire developing and gathering cycle. It starts with a seed being planted in the dirt — from the dirt planning, seeds rearing and water feed estimation — and it closes when robots get the gather deciding the readiness with the assistance of PC vision. The means associated with savvy cultivating utilizing brilliant innovations are demonstrated in Fig.1.3.

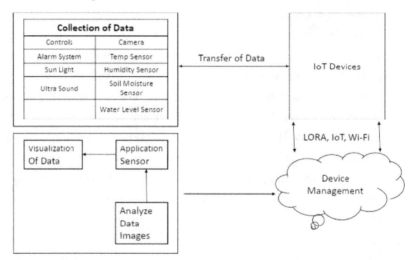

Figure 1.3 Steps involved in smart farming

1.2 Classification of Smart-Farming in Agriculture

1.2.1 Species Breeding

Species selection is a tedious process of searching for specific genes that determine the effectiveness of water and nutrients use, adaptation to climate change, disease resistance, as well as nutrients content or a better taste. Machine learning, in particular, deep learning algorithms, take decades of field data to analyze crops performance in various climates and new characteristics developed in the process.

Based on this data they can build a probability model that would predict which genes will most likely contribute a beneficial trait to a plant [15].

1.2.2 Species Recognition

While the traditional human approach for plant classification would be to compare color and shape of leaves, machine learning can provide more accurate and faster results analyzing the leaf vein morphology which carries more information about the leaf properties [16].

1.2.3 Soil management

For specialists involved in agriculture, soil is a heterogeneous natural resource, with complex processes and vague mechanisms. Its temperature alone can give insights into the climate change effects on the regional yield. Machine learning algorithms study evaporation processes, soil moisture and temperature to understand the dynamics of ecosystems and the impingement in agriculture [17].

1.2.4 Water Management

Water the executives in horticulture impacts hydrological, climatological, and agronomical equilibrium. Up until now, the most evolved ML-based applications are associated with assessment of day by day, week after week, or month to month evapotranspiration taking into account a more compelling utilization of water system frameworks and forecast of day by day dew point temperature, which distinguishes expected climate wonders and gauge evapotranspiration and vanishing [18].

1.2.5 Yield Prediction

Yield forecast is perhaps the most significant and well known themes in exactness horticulture as it characterizes yield planning and assessment, coordinating of harvest supply with request, and harvest the executives. Cutting edge approaches have gone a long ways past basic expectation dependent on the chronicled information, yet join PC vision innovations to give information in a hurry and extensive multidimensional examination of harvests, climate, and monetary conditions to take advantage of the yield for ranchers and populace [19].

6

1.2.6 Harvest Quality

The precise location and characterization of yield quality attributes can build item cost and diminish squander. In examination with the human specialists, machines can utilize apparently good for nothing information and interconnections to uncover new characteristics assuming part in the general nature of the harvests and to recognize them [20].

1.2.7 Infection Detection

Both in outside and nursery conditions, the most generally utilized practice in nuisance and infectious prevention is to consistently splash pesticides over the editing region. To be viable, this methodology requires huge measures of pesticides which brings about a high monetary and huge natural expense. ML is utilized as a piece of the overall exactness horticulture the board, where agro-synthetics input is focused regarding time, place and influenced plants [21].

1.2.8 Weed Detection

Aside from infections, weeds are the main dangers to trim creation. The most concerning issue in weeds battling is that they are hard to identify and segregate from crops. PC vision and ML calculations can improve recognition and segregation of weeds effortlessly and with no ecological issues and results. In future, these advancements will drive robots that will obliterate weeds, limiting the requirement for herbicides [22].

1.2.9 Animals Production

Like yield the board, AI gives precise forecast and assessment of cultivating boundaries to upgrade the financial productivity of domesticated animal's creation frameworks, for example, steers and eggs creation. For instance, weight foreseeing frameworks can assess the future loads 150 days preceding the butcher day, permitting ranchers to change diets and conditions individually [23].

1.2.10 Creature Welfare

In present-day setting, the domesticated animals are progressively treated as food holders; however as animals who can be despondent and depleted of their life at a homestead. Creatures conduct classifiers can associate their biting signs to the need in eating routine changes and by their development designs, including standing, moving, taking care of, and drinking, they can tell the measure of pressure the creature is presented to and foresee its helplessness to illnesses, weight gain and creation [24]. Smart farm framework is given in Fig.1.4.

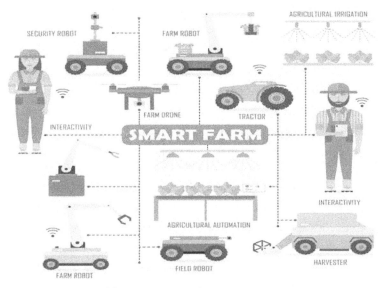

Figure 1.4 Concept of smart farm

1.3   Literature Review

The writing audit shows that the most well-known models in horticulture are Artificial and Deep Neural Networks (ANNs and DL) and Support Vector Machines (SVMs). ANNs are motivated by the human cerebrum usefulness and address an improved on model of the design of the natural neural organization imitating complex capacities, for example, design age, discernment, learning, and dynamic. Such models are regularly utilized for relapse and characterization undertakings which demonstrate their handiness in yield the board and recognition of weeds, illnesses, or explicit qualities [25]. The new advancement of ANNs into profound discovering that has extended the extent of ANN application altogether areas, including farming [26]. SVMs are paired classifiers that build a straight isolating

8

hyper plane to characterize information occurrences. SVMs are utilized for grouping, relapse, and bunching. In cultivating, they are utilized to foresee yield and nature of harvests just as domesticated animal's creation. More complicated errands, for example, creature government assistance estimation, require various methodologies, for example, different classifier frameworks in outfit learning or Bayesian models — probabilistic graphical models in which the investigation is attempted inside the setting of Bayesian induction. Despite the fact that still in the start of its excursion, ML-driven ranches are as of now developing into man-made consciousness frameworks [27]. As of now, AI arrangements tackle singular issues, yet with additional combination of robotized information recording, information investigation, AI, and dynamic into an interconnected framework, cultivating practices would change into with the purported information based agribusiness that would have the option to build creation levels and items quality [28].

In the modern synopsis of the industrial revolution, 4.0 where we have a limited amount of resources and their proper utilization is a subject of great concern, whether it's the utilization of water or utilization of minerals from ores all this indirectly affects our lives. With the limited availability of resources and increased consumption there prices have been rising up and so there sustainable utilization is necessary. Similarly, in the case of Farming where we need to feed a large number of customers, any kind of loss at any stage proves to be a huge loss to the economy and the user as well. Moreover, there is a lack of research data in this field. The main motive is to bring IoT and Machine Applied Farming to India, to ample up the technical application of AI and Machine Learning among Farmers, Researchers, and Government [29]. Throughout the long term, remote sensor organizations (WSNs) has been conveyed for keen agribusiness and food creation with an emphasis on natural observing, accuracy horticulture, machine and cycle control computerization and recognizability [30]. The capacity of WSN to self-sort out, self-arrange, self-analysis, and self-mend has settled on it a decent decision for keen farming and the food business [31]. The WSN is a framework that involves radio recurrence (RF) handsets, sensors, microcontrollers and force sources [32]. Nonetheless, with the crisis of IoT there has been a change in outlook from the utilization of WSN for shrewd farming to IoT as the significant driver of brilliant agribusiness [33]. The IoT incorporates a few advances that as of now exist, for example, WSN, RF recognizable proof, distributed computing, middleware frameworks and end-client applications. The utilization of IoT in horticulture is tied in with enabling ranchers with the choice devices and mechanization technologies that flawlessly incorporate items, information and administrations for better profitability, quality, and benefit. Late reviews on the IoT in agribusiness have zeroed in on the difficulties and limitations for huge scope pilots in whole production network in the agri-food area [34]. A portion of the major questions tended to are the requirement for new plans of action, security and protection, and information administration and possession arrangement. Other related overview on shrewd horticulture have generally centered on the utilization of WSNs [35]. While these review papers manage the utilization of sensor innovation and difficulties in the use of IoT to the food production network, the correspondence innovation were restricted to an ordinary strategy which utilizes low reach correspondence advances [36]. In harvest cultivating, there are a few natural factors that influence ranch produce. Gaining such information help to comprehend the examples and interaction of the homestead. Such information incorporates, the measure of precipitation, leaf wetness, temperature, dampness, soil dampness, saltiness, environment, dry circle, sun powered radiation, bother development, human exercises, and so forth [37]. The acquisition of such point by point record empowers ideal dynamic to improve the nature of the ranch produce, limit hazard, and expand benefits. For example, the sun based radiation information gives data about the plants openness to daylight from, where the rancher can recognize if the plants are appropriately uncovered or over uncovered [38]. The dirt dampness content gives data on the sogginess of the dirt which can help in controlling soil conditions and decrease the danger of plant infections. Besides, opportune and exact climate estimating information, for example, climatic changes and precipitation, can improve the efficiency level. Likewise, such information can help ranchers in the arranging organize and decrease the expense of work. The ranchers can likewise take remedial and preventive measures ahead of time dependent on the information gave. The bug development information can be gathered and distantly took care of live to the ranchers for bother control or used to give counsel to the ranchers dependent on record following of bug assaults [39]. A following and following framework ought to fundamentally include: data input, stockpiling, move, interaction, and yield. The info data incorporates the information of the whole life pattern of the item, the topographical inception, the current position, objective, and the partners engaged with the whole inventory network [40]. The frameworks ought to likewise incorporate memory to store the data throughout some stretch of time for innovative work purposes. The data move alludes to the way toward bringing together and normalizing the whole data. The following and following framework ought to likewise have the option to handle the information gathered lastly yield it to everybody required along the inventory network. The utilization of RFID in following from the creation stage, handling, transportation, stockpiling, appropriation, deals and after deals administrations is featured. It gives the capacity to gather, store, and break down information over a significant distance at a snappy speed [41].

Precision Agricultural Management System (PAMS) is based on IOT and web-GIS. PAMS have four architectures which are the spatial information platform, the IOT infrastructure platform, the Agricultural Management system and the Mobile client. The key techniques of PAMS are IOT, web-GIS, ICT, and location based service. The main function of PAMS is monitoring and management of agricultural farms. It analyses the data collected and gives suggestions to the staff about what to do at next stage. It reduces monitoring time and promote the level of management on the farm. It plays a major role in development of agriculture in china [42]. The objective of Intelligent Agricultural Management Information System (AMIS) based on IOT proposal is to apply Agricultural Information Technology [AIT] to every aspect of agriculture and has become the most effective means and tools for enhancing agricultural productivity and for making use of full agricultural resources [43]. It makes the digitization of each process in every aspect of agriculture and unites all the collected data. The purpose of AMIS is to improve the level of agricultural information process and enhance the intelligent management and decision of agricultural production [44]. The three layers of AMIS are data collection using RFID and provides processing information from farm gate to

11

restaurant plates, data transmission using GSM and wireless sensor networks (WSN) and data processing. E- Agricultural concepts proposed that the usage of ICT makes agriculture more efficient. It helps in both the product efficiency and process efficiency by means of reducing the cost and time in the functionalities involved in agriculture [45]. E-agriculture is an emerging field in the intersection of agricultural informatics, agricultural development and entrepreneurship, referring to agricultural services, technology, technology dissemination and information delivered or enhanced through the internet and related technologies [46]. The objective of his study was to increase agricultural productivity and decrease the poverty by introducing the new technologies to enhance agricultural processes and to educate rural communities about the importance of introducing new technologies in their production process. The ICT tools in this sector are Geographical Information System (GIS), Community Radio Stations, Internet and Web based Applications and Global Positioning System (GPS). The role of ICT is to monitor, processing and provide suggestions for pre-cultivation, crop cultivation and harvesting and post-harvest [47].The basic idea of the IOT is that virtually every physical thing in this world can also become a computer that is connected to the Internet [48]. The sectors in agriculture where IOT is used are Crop water management, Precision Management, Integrated Pest management and control and Food creation and supply [49]. The framework engineering in Fig. 1.1 the association between utilized segments. The circuit comprises of Arduino ATMEGA 2560, ESP 8266 Wi-Fi module, GPS sensor, PIR sensor, Soil dampness sensor and DHT 11/22 Temperature and Humidity sensor And I2C LCD screen [50]. The Precision Agriculture model is a result of the brisk headways in the Internet of Things and conveyed registering guidelines, which incorporate setting care and consistent events [51] present surveys about insightful farm undertakings, while multidisciplinary models abusing IoT sensors are assessed in progress of [52]. This [53]] use green-house gas assessment to checking the oil palm bequest used in the making of biodiesel, while [54] propose an expert structure to help farmers with choosing tomato collections planning limits or tendencies using cushioned reasoning on components like height, assurance from ailments, natural item size, natural item shape, yield potential, advancement, and normal item tone. Made by [55] uses IoT territories of interest as markers of forest

12

area fires in a locale where progressive instances of occasions can be removed from a dataset; [56] uses far off sensor association (WSN) advancement to screen a honey bee safe-haven settlement and assemble key information about activity/environment, while the makers of [57] present courses of action that can be composed into drones using Raspberry Pi module for improvement of reap quality in agricultural field. Major agri-business associations, that is, [58], which put gigantic resources in investigation and progression; pondering the normal viability, it achieves significant the insightful exhibiting used to regulate crop disillusionment risk and to help feed profitability in creatures creation presented in the composing [59]. In [60] develop a noticing structure that recognizes grape ailments in their starting stages, using factors, for instance, temperature, relative sogginess, soddenness, and leaf wetness sensor, while [61] uses an IoT device with an AI estimation that predicts natural conditions for infectious disclosure and expectation, using conditions, for instance, air temperature, relative air dampness, wind speed, and deluge fall; likewise, a system for area and control of diseases on cotton leaf close by soil quality checking is presented [62]. Rural Bridge is an IoT-based structure that uses sensors to accumulate legitimate information, for instance, soil soddenness level, soil pH regard, ground water level (GWL), and surface water level (SWL) for a keen and co-employable developing in the composing [63] present distant identifying used in nursery agriculture to grow the yield and giving characteristic developing. A SmartAgri-Food applied designing is proposed in [64], while the makers of [65] present web applications in the agri-food space; in [66] proposes the assessment to both the expansion and the relationship of estate creation rules. [67] makes sharp water-sharing strategies in semi-very dry zones; In [68] present plant diseases affirmation using authentic models; and, likewise, in [69] there are splendid hydroponics structures that abuse allowance in Bayesian associations. In [70] propose and plan a Persuasive Technology to enable splendid developing, while similarly abusing valid time-course of action for creation quality affirmation [71], in light of the fact that nowadays purchasers are stressed over sterilization confirmation related to prosperity and success. In [72], there is a structure that screens the cultivating field through Raspberry pi camera, allowing modified water framework subject to temperature, tenacity, and soil sogginess. In [73] generally revolve around in situ examination of the leaf zone record (LAI), a fundamental yield limit for insightful developing, while examinations of [74] present an IoT application, named 'AGRO-TECH', that is accessible by farmers to screen soil, gather, and water, which is moreover evolved by the makers of [75] develop an IoT-based precision developing procedure for exceptional yield groundnut agronomy proposing water framework timings and ideal utilization of fertilizers in regards to soil features. Emerging economies are also researching these models; the Government of China has performed investigation to save water for water framework gauging environment conditions [76], in like manner considering the earth decency and the air quality [77], while in [78] the canny farm perspective is proposed as an opportunity. Finally, an additional issue to take into accounts is data headway in the sending of a veritable application where data openness increase as time passes by [79]. [BLUE]. The stage associated with keen cultivating toward agribusiness 5.0 is given in Fig.1.5.

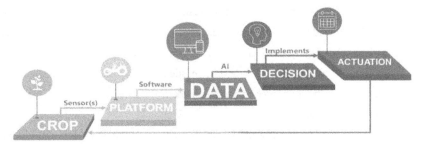

Figure 1.5 Process of Agriculture 5.0

1.4   Objective

The following time of Smart Computing will be situated completely on the execution of unavoidable processing associating the actual world with the cloud. Two advances will help in the usage Machine Learning and Internet-of-Things. These two advances are integral and frequently add to the effectiveness of the frameworks being planned yet at the same time are totally unique. Web of Things is an innovation where various articles are given abilities of giving data and making moves according to the pre-modified information conditions, the program is frequently into the regulator. AI is tied in with preparing shrewd machines equipped for gaining from past/authentic information and dependent on that experience and

14

pre-taken care of calculations settle on choices later on. In our examination, we are wanting to plan frameworks and answers for improve crop efficiency in dry locales, for example, those found in Rajasthan, Gujarat, and different spots on the planet. The issues here are novel as in these zones frequently have restricted water for water system and ranchers are generally reliant on downpours which regularly play no-show prompting hefty misfortunes and low efficiency of the land. By utilizing appropriate IoT and Machine learning methods we will attempt to improve efficiency by giving standard contributions to the rancher concerned. By utilizing IoT equipment and programming mix, we will continually screen little insights about the wellbeing of yields by estimating the temperature, stickiness, dampness, pH, and a few different boundaries of the dirt and climate. AI models will be utilized to recognize sicknesses in the yield and propose applicable measures to defeat these issues. By utilizing this strategy if the rancher goes over any circumstance where the yield is tainted by particular kinds of plant illnesses, at that point Machine Learning will be utilized to recognize the sickness and afterward plan the further procedures as needs be. This won't just expand crop profitability yet in addition diminish the expense of mediations, for example, pesticides and insect poisons which can be chosen with the assistance of models made by us. The objectives of the projected work are as follows:

1. Employment of artificial intelligence in agriculture for optimization of irrigation and application of pesticides and herbicides.
2. A comprehensive evaluation on computerization in agriculture using artificial intelligence.
3. Comprehensive study of Machine Learning applications in IoT based agriculture and Smart farming.

As of now, Artificial Intelligence and IoT innovations are being utilized in created countries for Industrial creation essentially to distinguish the quality issues in the item and furthermore to improve the efficiencies of machines delivering items. These have demonstrated to be extremely viable and are useful however the expense of usage includes high capex ventures at the start. These innovations are creating at a high speed and it is seen that as the selection rate builds, the general expense of usage will diminish. One more significant issue within reach is the mindfulness level of the ranchers and prodding them to change the regular strategies for cultivating and embracing the most recent innovation. The outlook change is a major test which should be done to influence huge scope selection of the more up to date innovations to take care of Indian 125+ Crore populace with the zone under development diminishing each year, expanding profitability from existing area banks under development is the lone decision [80]. As the name recommends, the Internet of Things is an organization of things. Prior all the things which are clearly non-living in nature were being worked by people physically and the articles were not liable for the yield of the choices made by people to work. Yet, in this advanced world all the items which can likewise be alluded to as things are associated with the web or basically any organization. The fundamental explanation behind interfacing these articles inside an organization is to work the items by a Controller. A Controller is a gadget which is made to control actuators in reality dependent on the given conditions and circumstances which are as of now predefined for the regulator. In light of these conditions and circumstances the regulator will choose in the simple future on the bases of the guidelines to control all the gadgets or the items which are being associated with the regulator either through wired medium or remote medium. This way the regulator is given the obligation to control the actual gadgets. Coming to Machine Learning innovation, ML is a field of study wherein the PC is prepared so that the PC gets a capacity to learn with no express programming. AI and Internet of Things both the innovation can be incorporated so that together they can mechanize the current technique for work and even gain from the experience of the work to utilize the involvement with the future work appointed. Our fundamental point here is to utilize these both the innovation to make the cultivating simple more cycle organized and even use the past experience to sidestep various future perilous conditions. Developing is an indispensable piece of human perseverance and by using these front line advances to improve developing is the best way out. Web of Things will be used to check sogginess in the soil, pungency in the earth, pH level of the earth. These assessments will be continually assessed by using the sensors as actuators and thereafter these assessments will be passed to the controller which is truly controlling the sensors. These Measurements will be valuable in checking the

16

earth where the harvests are created. We will moreover screen the environment by taking various assessments like temperature of the environment, tenacity in the environment and clamminess in the environment [81]. This will help us with anticipating the advancement of the yields depending on the climate ideal for the reap. Computer based intelligence will moreover accept a huge part in guarding the yields from a wide scope of gather sickness and will in like manner perceive the contamination so we can regard the reap as necessities be with authentic measures and care. For perceiving the ailment we will at first deal with past data of plants and their disease with photos to set up our estimation once the arrangement of the computation is done the count is as of now set up to recognize the disorder. What we need to do is basically give a nice photo, rest all that will be figured out by the real estimation [82]. The IoT based keen cultivating is given in Fig.1.6.

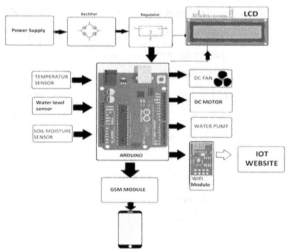

Figure 1.6 IOT based smart farming

1.5   Conclusion

These days, we experience a wealth of Internet-of-Things (IoT) middleware arrangements that give network to sensors and actuators to the Internet. To acquire an inescapable selection, these middleware arrangements, alluded to as stages, need to meet the assumptions for various parts in the IoT biological system, including gadget suppliers, application engineers, and end-clients, among others. Assessments

17

are done in different exploration papers for these stages, both exclusive and open-source, based on their capacity to meet the biological system assumptions and a hole is addressed of the ebb and flow IoT and AI/ML scene regarding (I) the help for execution of AI/ML for crop infection recognition, (ii) heterogeneous detecting and activating advances, (iii) Absence of recorded information to prepare Machine learning models for illness ID, (iv) the information proprietorship and its suggestions for security and protection, (v) information handling and information sharing abilities, (vi) web network and practicality of utilizing new age RF advances in distant, (vii) the fulfillment of an IoT environment, The hole investigation plans to feature the insufficiencies of the present answers for improve their mix to the upcoming environments. The assessment ought to be more centered on how conventional methods and data can be installed into the advanced universe of AI and ML to get the advantage of both the universes old and new and refine land richness and yield.

Chapter 2
A STUDY ON AGRICULTURE USING ARTIFICIAL-INTELLIGENCE

2.1   Introduction

The world's population is assumed to be nearly 10 billion by 2050, boosting agricultural order-in a situation of humble financial development by somewhere in the range of 50% contrasted with 2013 [81]. At present, about 37.7% of total land surface is used for crop production. From employment generation to contribution to National Income, agriculture is important. It is contributing a significant portion in the economic prosperity of the developed nations and is playing an active part in the economy of the developing countries as well. The augmentation of agriculture has resulted in a significant increase in the per-capita income of the rural community. Thus, placing a greater emphasis on agricultural sector will be rational and apposite. For countries, like India, the agricultural sector accounts for 18% of GDP and provides employment to 50% of the country's workforce. Development in the agricultural sector will boost the rural development, further leading toward rural transformation and eventually resulting in the structural transformation [82]. With the advent of technology, there has been observed a dramatic transformation in many of the industries across the globe [83]. Surprisingly, agriculture, though being the least digitized, has seen momentum for the development and commercialization of agricultural technologies [84]. Artificial Intelligence (AI) has begun to play a major role in daily lives, extending our perceptions and ability to modify the environment around us [85]. The authors [86] gave a method for harvest planning based on the coupling of crop assignment with vehicle routing is presented [87]. With this emerging technologies the workforce which were restricted to only a minimal industrial sectors are now contributing to numerous sectors [88]. AI is based on the vast domains like Biology, Linguistics, Computer Science, Mathematics, Psychology and engineering. The authors presented a brief overview of the current implementation of agricultural automation [89]. The paper also addresses a proposed system for flower and leaf identification and watering using IOT to be implemented in the botanical farm [90]. The basic concept of AI to develop a technology which functions like a human brain [91] This technology is perpetrated by studying how human brain thinks, how humans learn, make decisions, and work while solving a problem, and on this ground intelligent software and systems are developed. These software are fed with training data and further these intelligent devices provide us with desired output for every valid input, just like the human brain. Vast domains including Machine Learning and Deep learning are core part of AI [92]. While AI is the science of making intelligent machines and programs [93], ML is the ability to learn something without being explicitly programmed and DL is the learning of deep neural networks [94]. The main subjective of AI is to make problem solving facile which may include the use of ANN [95]. ANN is a processing algorithm or a hardware whose functioning is inspired by the design and functioning of a human brain [96]. Neural networks have a remarkable ability of self-organization, and adaptive learning. It has replaced many traditional methods in numerous fields like Computer Science, Mathematics, Physics, Engineering image/signal processing, Economic/ Finance, Philosophy, Linguistics, Neurology. ANN undergoes the process of learning. Learning is the process of adapting the change in itself as and when there is a change in environment. There are two learning techniques, supervised learning and unsupervised learning. The authors, encloses the connected relations between the various embedded systems and the AI technology coherent with the agricultural field, it gave a brief about the various applications of neural networks, ML in this sector for precision farming [97]. AI is an emerging technology in the field of agriculture. AI-based equipment and machines, has taken today's agriculture system to a different level. This technology has enhanced crop production and improved real-time monitoring, harvesting, processing and marketing [98]. The latest technologies of automated systems using agricultural robots and drones have made a tremendous contribution in the agro-based sector. Various hi-tech computer based systems are designed to determine various important parameters like weed detection, yield detection and crop quality and many other techniques [99]. This paper encompasses the technologies used for the automated irrigation, weeding and spraying to enhance the productivity and reduce the work load on the farmers. Various automated soil sensing techniques are discussed [100]. In [101], brought together temperature and moisture sensors to close the loop holes of the vehicle predictions. The robots used in sensing were localized by GPS modules and the location of these robots was tracked using the Google maps. The data from the robots was fetched through Zigbee wireless protocol. The readings were displayed on the 16 × 2 LCD display which was integrated to the LPC2148 microcontroller. The latest automated weeding techniques are discussed and the implementation of drones for the purpose of spraying in the fields is discussed followed by the types of sprayers utilized on UAVs. Further speaking about drones, yield mapping and monitoring is discussed beginning with the an outline of the yield mapping processes followed by the programming of the software and briefing about the calculation as well as calibration process. Finally the processing of these yield maps is illuminated.

The advances which are AI-put together help to improve capability with respect to the entire the fields and furthermore deal with the troubles looked by changed organizations recollecting the various fields for the agrarian zone like the collect yield, water framework, soil content recognizing, crop-checking, weeding, crop establishment [102]. Agricultural robots are verifiable solicitation to pass on high regarded utilization of AI in the referred to territory. With the overall people removing, the cultivating region is defying a crisis, anyway AI can pass on really vital plan. Man-made knowledge based imaginative courses of action has enabled the farmers to convey more yield with less data and even improved the idea of yield, also ensuring snappier go-to-publicize for the yielded crops. By 2020, farmers will use 75 million related devices. By 2050, the ordinary farm is needed to make a typical of 4.1 million data concentrates every day. The various habits by which AI has contributed in the cultivating zone are according to the accompanying:

2.2   Various Topologies in AI based farming

2.2.1   Picture affirmation and wisdom

In [103], it was said that lately, an extending interest has been found in free UAVs and their applications including affirmation and observation, human body area and geo-constrainment, search and rescue, forest area fire disclosure [104]. Because of their adaptability similarly as dumbfounding imaging development

21

which covers from transport to photography, the ability to be guided with a far-away controller [105] and the devices being competent in air which engages us to do a ton with these contraptions, robots or UAVs are getting dynamically standard to show up at mind boggling heights and eliminates and finishing a couple of utilizations [106].

2.2.2 Capacities and workforce

In [107], it is seen that man-made thinking makes it serviceable for farmers to gather enormous proportion of data from government similarly as open destinations, take apart each and every bit of it and outfit farmers with answers for some questionable issues similarly as it gives us a more clever strategy for water framework which achieves better re-visitation of the farmers. Due to man-made intellectual competence, developing will be found to be a mix of inventive similarly as natural capacities in the near future which will not simply fill in as an unrivaled outcome in the matter of significant worth for all the farmers yet furthermore limit their setbacks and remaining weights. UN communicates that, by 2050, 2/third of absolute people will be living in metropolitan districts which arises a need to reduce the load on the farmers. PC based insight in agribusiness can be applied which would motorize a couple of cycles, reduce possibilities and give farmers an almost straightforward and capable developing.

2.2.3 Grow the yield

In [108], it was seen that grouping decision and seed quality set the most outrageous execution level for all plants. The emerging headways have helped the best assurance of the harvests and even have improved the decision of cream seed choices which are generally suitable for farmer's necessities. It has executed by perceiving how the seeds react to various environment conditions, unmistakable soil types. By get-together this information, the chances of plant contaminations are diminished. As of now we can meet the market designs, yearly outcomes, buyer needs, likewise farmers are successfully prepared to extend the benefit from crops.

2.2.4 Chat-bots for farmers

Chat-bots are just the conversational far off partners who mechanize relationship with end customers. Electronic thinking powered chatbots, close by AI methodology has enabled us to appreciate trademark language and work together with customers in away more redid way. They are fundamentally ready for retail, travel, media, and agriculture has used this office by assisting the farmers with finding solutions to their unanswered requests, for offering direction to the mind giving various recommendations moreover [109].

2.3 Advancement in Smart Farming

Advanced mechanics and Autonomous Systems (AMAS) are presented in huge areas of the economy with moderately low profitability, for example, agri-Food. As indicated by UK-RAS White papers (2018) the UK agri-Food chain, from essential cultivating through to retail, produces over £108bn p.a., and with 3.7memployees in a genuinely global industry yielding £20bn of fares in 2016. Mechanical technology has assumed a considerable part in the rural creation and the executives. The scientists have now begun underscoring on innovations to plan independent horticultural devices as the traditional cultivating apparatuses needed effectiveness [110]. The primary motivation behind thinking of this innovation is to supplant human work and produce powerful advantages on little just as huge scope creations [111]. In this area, the space for automated advancements has intensified efficiency monstrously [112]. The robots are performing different agrarian activities independently, for example, weeding, water system, guarding the ranches for conveying successful reports, guaranteeing that the unfriendly ecological conditions don't influence the creation, increment accuracy, and oversee singular plants in different new manners. Coming up with such an innovation accompanied the presentation of a machine called Eli Whitney's cotton gin. It was concocted in 1794 by U.S. - conceived innovator Eli Whitney (1765–1825), a gadget which changed cotton creation by altogether quickening the way toward removing seed from cotton fiber. It made 50 pounds of cotton in one day. Hence this brought forth the self-sufficient agrarian robots. An essential robotized model was acquainted with decide the real situation of seeds [113]. Ultrahigh exactness arrangement of seed was additionally settled. Components that guarantee that the seed planted has zero ground speed [114]. This is significant as it guarantees that the seed doesn't ricochet from its real situation after the dirt effect. The status or the improvement of plant was recorded via computerized machines. Different biosensors were set up to screen the plant development and furthermore to distinguish plant sicknesses [115].The interaction of manual weeding was supplanted by the laser weeding innovation, where a portable shone infra-red light upsets the cells of the weeds, this shaft was constrained by PCs [116]. For the successful utilization of water, computerized water system frameworks were additionally settled. Various types of components are

2.3.1 Irrigation

The horticulture area burns-through 85% of the accessible freshwater assets across the world. What's more, this rate is expanding quickly with the populace development and with the increment in food interest. This leaves us with the need to think of more effective innovations to guarantee appropriate utilization of water assets in water system. The manual water system which depended on soil water estimation was supplanted via programmed water system booking methods. The plant evapotranspiration which was subject to different climatic boundaries, for example, stickiness, the breeze speed, sun oriented radiations and even the harvest factors, for example, the phase of development, plant thickness, the dirt properties, and nuisance was thought about while actualizing self-governing water system machines. It is examine about the distinctive water system strategies with the essential rationale of building up a framework with diminished asset utilization and expanded proficiency. Gadgets like ripeness meter and PH meter are set up on the field to decide the fruitfulness of the dirt by recognizing the level of the essential elements of the dirt like potassium, phosphorous, nitrogen [117]. Programmed plant irrigators are planted on the field through remote innovation for trickle water system. This technique guarantees the richness of the dirt and guarantees the powerful utilization of water asset. The innovation of keen water system is created to expand the creation without the inclusion of huge number of labor by distinguishing the degree of water, temperature of the dirt, supplement substance and climate anticipating. The activation is performed by the microcontroller by turning ON/OF the irrigator siphon. The M2M that is, Machine to Machine innovation is been created to facilitate the correspondence and information dividing between one another and to the worker or the cloud through the primary organization between all the hubs of the agrarian field [118]. In [119] a mechanized automated model was produced for the location of the dampness substance and temperature of the Arduino and Raspberry pi3. The information is detected at ordinary stretches and is shipped off the microcontroller of Arduino (which has an edge level equipment associated with it), it further believers the info simple to advance. The sign is shipped off the Raspberry pi3 (inserted with KNN calculation) and it imparts the sign to Arduino to begin the water hotspot for water system. The water will be provided by the asset as shown by the essential and it will moreover invigorate and store the sensor regards. In [120], an electronic water framework was also advanced with the development of Arduino for lessening the work and time usage during the time spent water framework. In [121], the chance of successful and motorized water framework structure by making distant sensors using the advancement of Arduino which can develop the creation to 40%. Another structure for mechanized water framework was given in [122]. In this strategy different sensors were worked for different purposes like the earth sogginess sensor to perceive the moistness content in the soil, the temperature sensor to recognize the temperature, the propel regulator sensor to keep up squeezing factor and the sub-nuclear sensor for better collect turn of events. The foundation of cutting edge cameras. The yield of all of these contraptions is changed over to modernized sign and it is transported off the multiplexer through distant association, for instance, Zigbee and territory of interest. The primary method was the subsurface spill water framework measure, which restricted the proportion of water hardship in view of scattering and overflow as it is clearly covered under the reap. Later experts went with different sensors which were used to perceive the need of water supply to the fields as soil moistness sensor and deluge drop sensor, which were told through distant broadband association and filled by daylight based sheets. The deluge drop sensor and soil moistness sensor exhorts the farmer about the clamminess content in the earth through SMS in their cell using GSM module. As necessities be the farmer can give orders using SMS to ON and OFF the water supply. Thusly we can consider that this system will perceive part or area in the fields which required more water and could hold off the farmer from diluting when it's coming [123]. Soil moistness sensors use one of only a handful few advancements used to measure the earth sogginess content. It is covered near

25

the root zones of the harvests [124]. The sensors help in decisively choosing the soddenness level and impart this scrutinizing to the controller for water framework. Soil moistness sensors in like manner help in basically directing water [125]. One strategy for clamminess sensors is the water on revenue water framework in which we set the edge as demonstrated by the earth's field limit and these sensors permits your controller to water exactly when required. Right when the booked time appears, the sensor scrutinizes the moistness substance or level for that particular zone, and watering will be allowed in that zone just if the soddenness content is under the cutoff. The various was the suspended cycle water framework which requires water framework length not under any condition like the water on interest water framework. It requires the starting time and the term for each zone [126].

2.3.1.1 Dielectric strategy

The dampness in the dirt is determined by the sensors which fundamentally assess the dampness content in the dirt dependent on the dielectric consistent (soil mass permittivity) of the dirt. The measure of water system required can likewise be resolved based on the dielectric steady [127]. Kinfolk [128], projected a mechanized framework that utilizes dielectric soil dampness sensors for continuous water system control. The estimation technique dependent on the dielectric properties is viewed as the most potential one [129]. The data with respect to soil types influence the exactness to dielectric dampness sensors is was examined [130]. The dielectric consistent is just the limit of soil to move force or power. The dirt is involved different parts like minerals, air and water, consequently the assessment of its dielectric reliable is controlled by the overall responsibility of all of these sections. Since the assessment of the dielectric estimation of water is significantly greater than the assessment of this steady for the other soil parts, the assessed estimation of permittivity is fundamentally addressed by the closeness of dampness in the dirt. One strategy to ascertain the connection between the dielectric consistent and volumetric soil dampness is the condition of [131]:$Mositure = -5.2 * 10^{-3} + 3.32 * 10^{-2} - 4.4 * 10^{-5} + 53 * 10^{-6}$

The other method used for determining the dielectric constant is the by the Time Domain Reflectometry (TDR). It is determined on the basis of the time taken by an electromagnetic wave to propagate along a transmission line that is surrounded by the soil. As we probably are aware, the propagation velocity is an element of the dielectric constant, therefore it is legitimately corresponding to the square of the transmission time down and back along the transmission line [132]

2.3.1.2 Neutron moderation

This is another technique for deciding the moisture content in the soil. In this strategy fast neutrons are launched out from a decomposing radio dynamic source like 241Am/9Be and when these neutrons slam into particles having a similar mass as theirs(protons, H+), they drastically slow down, making a "cloud" of "thermalized" neutrons. As we already know that water is the primary wellspring of hydrogen in soil, the thickness of thermalized neutrons around the test is about corresponding to the division of water present in the soil. The arrangement of the test is as a long and limited chamber, comprising of a source and a finder [133]. The estimations are taken in this test by bringing the test into an entrance tube, which is as of now presented in the soil. One can decide soil amount of moisture in the soil at various profundities by balancing the test in the cylinder at various profundities. The moisture substance is gotten with the assistance of this gadget dependent on a direct alignment between the check pace of thermalized neutrons read from the test, and the soil moisture substance got from adjacent field tests. The installation of sensors plays an important role in the efficient implementation of irrigation robotics. One can use a single sensor to control the irrigation of multiple zones in the fields. And one can also set multiple sensors to irrigate individual zones. In the first case where one sensor is utilized for irrigating multiple zones, the sensor is places in the zone which is the driest of all or we can say the zone which requires maximum irrigation in order to ensure adequate irrigation in the whole field. The placement of the sensors should be in the root zone of the crops (ensuring that there are no air gaps around the sensor) from where the crops extract water. This will ensure the adequate supply of water to the crops. Later, we need to connect the SMS controller with the sensor. The controller will control the working after the sensor responds. After making this connection the soil water threshold needs to be selected. Then water is applied to the area where the sensor is buried and it is left as it is for a day. The water content now is the threshold for the sensor for scheduled irrigation as described earlier. After fetching the data through the sensors the microcontrollers

27

come into work. It is the major component of the entire automated irrigation process. The whole circuit is supplied with power up to 5 V with the help of transformer, a bridge rectifier circuit (which is a part of electronic power supplies which rectifies AC input to DC output) and voltage regulator. Then the microcontroller is programmed. The microcontroller receives the signals from the sensors. The OP-AMP acts as an interface between the sensors and the microcontroller for transferring the sensed soil conditions. The irrigator pumps thus operates on the information of the soil properties at run time as indicated in Fig. 2.1 and Fig. 2.2. The irrigation process can therefore be automated with the help of moisture sensors and microcontrollers [134] given in Table 2.1.

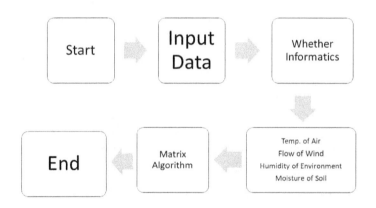

Figure 2.1 Irrigation Process

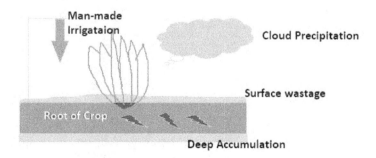

Figure 2.2 AI process in Irrigation

28

Table 2.1 Summary of Irrigation Automation Using Various Artificial Intelligence Technologies [135]

S.No.	Algorithms	Method of Evapotranspiration	Other Technologies	Advantages/ Results
1	PLSR as well as other regression algorithms.	Evapotranspiration model	Sensors for data collection, for the use of equipment.	More efficient and cost-effective.
2	Control system based on an artificial neuron network.	Evapotranspiration model	Sensors for measurement of soil, temperature, wind speed, etc.	Automation
3	Fuzzy Logic	FAO Penman-Monteith method		Optimization
4	ANN (multilayer neural model) Levenberg Marquardt, Backpropagation	Penman–Monteith method		Evaporation decreased due to schedule and savings observed in water and electrical energy
5	Fuzzy Logic		WSN, Zigbee	Experimental results verification. Can be applied to home gardens and grass
6	ANN Feed Forward, Back-propagation			Optimization of water resources in a smart farm.
7	Fuzzy Logic Controller	Penman–Monteith method	Wireless sensors	Drip irrigation prevents wastage of water and evaporation
8	Machine Learning algorithm		Sensors, Zigbee, Arduino microcontroller	Prediction and tackles drought situations

2.3.2 Weeding

In [55], it was expressed about an investigation on the opposition among plants. After his point by point research on the equivalent, he came reasoned that the opposition among the plants for water starts when their underlying foundations in the dirt cover to retain water and supplements and weeds were the most grounded contenders for water. The water necessity for the ethereal pieces of the plant is the quantity of pounds of water used to deliver a pound of dry matter. The wild mustard plant requires four fold the amount of water as an all around created oat plant, and the basic ragweed plant requires three fold the amount of water as a corn plant to arrive at development. One can compute the water necessity per section of land is dictated by increasing the creation of the plant in pounds of dry matter per section of land times the plant's water prerequisite. Light is likewise a fundamental segment for the development of the plants. Weeds which develop tall, for the most part obstructs the method of light to the plants. Sometimes weeds like green foxtail and redroot pigweed are intolerant of shade but may times weeds like field bindweed, common milkweed spotted spuroe, and Arkansas rose are shade tolerant. According to a study by researchers of the Indian Council for Agricultural Research, the country India, loses agricultural produce worth over \$11 billion — more than the Centre's budgetary allocation for agriculture for 2017–18 annually due to weeds. So to remove these weeds from the fields is of great importance otherwise it will not only occupy the land space but will also adversely affect the growth of other plants [136]. In [137], it was brought up a vision based weed detection technology in natural lighting. It was created utilizing hereditary calculation distinguishing a locale in Hue-Saturation-Intensity (HSI) shading space (GAHSI) for open air field weed detecting. It utilizes outrageous conditions like radiant and shady and these lightning conditions were mosaicked to discover the likelihood of utilizing GAHSI to find the locale or zones in the field in shading space when these two boundaries are displayed at the same time. They came about given by the GAHSI gave proof to the presence and severability of such a locale. The GAHSI execution was estimated by contrasting the GAHSI-portioned picture and an comparing hand sectioned reference picture. In this, the GAHSI achieved equivalent performance. Before developing a weed control automated system we need to differentiate between the crop seedlings

30

and the weeds [138]. A method was applied for recognition of carrot seedlings from those of ryegrass. In [139], implemented this method by the simple morphological characteristic measurement of leaf shape. This method has varying effectiveness mostly between 52 and 75% for discriminating between the plants and weeds, by determining the variation in size of the leaf. Another method for weeding was implemented using digital imaging. This idea involved a self-organizing neural network. But this method did not give appropriate results which were expected for commercial purposes, it was found that a NN based technology already existed which allows one to find the differences between species with an accuracy exceeding 75%. In the contemporary world many automated systems are developed given Table 2.2. But earlier various physical methods were used which relied on the physical interaction with the weeds. In [140], proposed that weeding depends on the position and the number of weeds. Classical spring or duck foot tines were used to perform intra row weeding by breaking the soil and the interface of roots by tillage and thus promote the witling of the weeds. But this is not advisable method as tillage can destruct the interface between the yield and the dirt. Along these lines, further no contact techniques like the laser medicines [141] and miniature showering, which don't influence the contact between the roots and the dirt was created. In [142], it clarified the technique for the utilization of agrarian robots for the concealment of weeds and creating strategies for controlling the stances of robots if there should arise an occurrence of lopsided fields in the rice development.

Table 2.2: Analysis of different applications

S. No.	Application	Crop	Algorithms for Weed Detection	Weed Removal Methods	Accuracy
1	Precision Weed Management	Pepper plants, artificial plants	Machine Vision, Artificial Intelligence	A smart sprayer	–
2	Autonomous Weeding Robot	Sugar beet	Machine vision algorithm	High power lasers for intra-row weeding proposed	–

3	Weeds Detection in Agricultural Fields	Data augmentation for image preprocessing	Convolutional neural networks for weed detection	Herbicide Spray	70.50%
4	Robot for weed control	Sugar Beets	Machine Vision	Rotatory hoe/ Mechanical removal	92%
5	Weeding Robot	Rice	–	Motion of robot prevents weed growth	–
6	Weed Prevention Robot	Rice	–	Motion of robot	–
7	Weed Detection	Sugarcane	Color Based and Texture Based algorithms; Greenness Identification; Fuzzy Real Time Classifier	Robotic arms for mechanical removal	92.90%
8	Weed Control System	Lettuce	Machine Vision	Electrical Discharge	84% (detection)
9	Robotic Weed Control	Cotton	Machine Vision algorithm based on Mathematical morphology	Chemical spraying	88.8% sprayed

It utilized the strategy for Laser Range Fielder (LRF) for stifling the weeds and controlling robot's stance. In [143], introduced an automated weed control framework. The robot was implanted with various vision frameworks. One was the dim level vision which was utilized in building up a column structure to manage the robot along the lines and the other vision was shading based which was generally significant and used to separate a solitary among the weeds. The line acknowledgment framework was created with a novel calculation with a precision of

±2 cm. The principal preliminary of this framework was actualized in a nursery for weed control inside a column of harvests. A similar innovation was referenced in the exploration done by [144]. The vision based innovations which were utilized to direct the robots along the line construction to eliminate weeds and to separate the single yield among the weed plants. The different weeding frameworks are:

2.3.2.1 Synthetic based

In this innovation, the framework comprised of 8 spouts at the back which were utilized for showering herbicides. The entire framework partitioned the pictures caught in 8 × 18 little square shapes or we can say impedes, every one of these squares covered a region of 8128 sq. mm. Afterward, each column which comprised of these squares relating to number of spouts was analyzed and prepared consistently. In the wake of analyzing the squares, each crate containing weeds are splashed. On can likewise isolate the pictures into 16 × 40 squares, for this situation each squares covers a territory of roughly 8768 sq. mm. Hence, for this situation we need 16 spouts rather than 8. The further handling, that is, the assignment of showering was done based on the conditions referenced. The conditions are:

1. On the off chance that the square analyzed comprising of weed pixels surpassing 10% of the absolute territory of the square, at that point it is arranged into a weed block.

2. All the squares analyzed are showered with herbicides.

3. At that point after these two conditions, the weeds whose territory equivalent to or over 30% is showered should be obliterated.

4. The herbicide which is showered in this technique is a particular herbicide, which annihilates just the weeds and not different plants. The initial two conditions referenced above characterizes the where the herbicides are to be splashed, that is, characterizes the regions which requires showering. The primary condition referenced diminishes the zones which contains extremely modest quantity of weeds and which doesn't need splashing. This is a significant piece of weeding. To obliterate weeds, all the pieces of the weeds don't need.

2.3.2.2. Handling yield maps

With the usage of a Geographic Information System (GIS) programming, the yield decided at each field zone can be appeared. The crude log report, contains centers which are recorded during turns and as the grain travel through a unite is a conceded cycle (except if continuous correction is associated), the sensor assessments disregard to contrast with the cautious assemble territories. To forgo these prominent mix-ups, the unrefined data is moved to compensate for the joining delay. Progressively completed, the centers which contrast with the header up position are cleared. Settings for grain stream deferment are join and a portion of the time even collect unequivocal, yet ordinary regards for grain yields reach out from around 10 to 12 s. A few concentrations at the beginning and around the completion of a pass should be ousted too [145]. These centers are implied as start and end-pass delays. Start pass delays happen when the grain stream has not offset considering the way that the lift is step by step finishing off yet the merge starts assembling the yield. In this manner, end-pass deferrals happen when the join moves out of the yield and grain stream continuously diminishes to zero when the lift is completely depleted. Moving of crude data to address for grain stream delay and avoidance of centers that address header status up and start and end-pass deferrals is the fundamental data isolating strategy fused with programming gave yield planning structures.

2.3.2.3 Challenges and future scope

Horticulture has been handling critical challenges like shortfall of water system framework, change in temperature, thickness of groundwater, food shortage and wastage and generously more. The destiny of developing relies generally upon gathering of different psychological arrangements. While enormous scale research is as yet in advancement and a few applications are now accessible on the lookout, the business is still exceptionally underserved [146]. With regards to taking care of practical difficulties looked by ranchers and utilizing self-sufficient dynamic and prescient answers for tackle them, cultivating is still at an incipient stage. To investigate the gigantic extent of AI in farming, applications should be more powerful [147]. Really at that time can it handle regular changes in outer conditions, encourage continuous dynamic and utilize fitting system/stage for gathering context oriented information in an effective way. Another significant perspective is the excessive expense of various psychological arrangements accessible on the lookout for cultivating. The arrangements need to turn out to be more moderate to guarantee that the innovation arrives at the majority. An open source stage would make the arrangements more reasonable, bringing about quick appropriation and higher entrance among the ranchers. The innovation will be valuable in aiding ranchers in high yielding and having a superior occasional harvest at normal span. Numerous nations, including India, the ranchers are subject to rainstorm for their development. They mostly rely upon the forecasts from different offices over the climate conditions, particularly for downpour took care of development. The AI innovation will be helpful to foresee the climate and different conditions identified with farming like land quality, groundwater, crop cycle, and irritation assault, and so forth the precise projection or forecast with the assistance of the AI innovation will diminish the majority of the worries of the ranchers. Simulated intelligence driven sensors are exceptionally valuable to extricate significant information identified with agribusiness. The information will be helpful in upgrading creation. In horticulture, there is an enormous extension for these sensors. Horticulture researcher can determine information like nature of the dirt, climate and groundwater level, and so forth; these will be helpful to improve the development cycle. Simulated intelligence enabled sensors can likewise be introduced in the automated collecting gear to get the information. It is conjectured that AI-based warnings would be valuable to expand creation by 30%. The greatest test to cultivating is the yield harm because of any sort of catastrophes including the nuisance assault. More often than not because of absence of the legitimate data ranchers lose their harvests. In this digital age, the innovation would be valuable for the ranchers to shield their development from any sort of assaults. Artificial intelligence empowered picture acknowledgment will be valuable toward this path. Numerous organizations have actualized robots to screen the creation and to distinguish any sort of nuisance assaults. Such exercises have been fruitful ordinarily, which gives the motivation to have a framework to screen and secure harvests. An automated focal point focuses in on the yellow blossom of a tomato seedling. Pictures of the plant stream into a computerized reasoning calculation that predicts definitely what amount of time it will require for the bloom to turn into a ready tomato prepared for picking, pressing, and the produce part of a supermarket. The innovation is being created and explored at Nature Fresh Farms, a

20-year-old organization developing vegetables on 185 sections of land among Ontario and Ohio. Realizing precisely the number of tomatoes will be accessible to sell later on makes the work of the outreach group simpler and straightforwardly benefits the primary concern, said Keith Bradley, IT Manager for Nature Fresh Farms. It's just a single illustration of AI changing farming, an arising pattern that will help spike an agrarian upheaval. From recognizing irritations to foreseeing what yields will convey the best returns, man-made brainpower can assist humankind with defying perhaps the greatest test: taking care of an extra 2 billion individuals by 2050, even as environmental change disturbs developing seasons, transforms arable land into deserts, and floods once-prolific deltas with seawater. The United Nations gauges we should build food creation half by the center of the century. Farming creation significantly increased somewhere in the range of 1960 and 2015 as the total populace developed from 3 billion individuals to 7 billion. While innovation assumed a part as pesticides, manures, and machines, a significant part of the additions can be credited to just furrowing more land—cutting backwoods and redirecting new water to fields, plantations, and rice paddies. We should be more creative this time around. Man-made intelligence is probably going to change horticulture and the market in the following not many years. The innovation has been helpful for the ranchers to comprehend different kinds of mixture developments which would yield them more pay inside the restricted time period. The legitimate execution of AI in farming will help the development cycle and to make a vibe for the market. According to the information with driving establishments, there is a colossal wastage of the food across the world and utilizing the correct calculations, this issue can likewise be tended to which won't just set aside the time and cash yet it will prompt practical turn of events. There are better possibilities for computerized change in horticulture upheld by utilizing advances like AI. However, everything relies upon the tremendous information which is very hard to assemble in light of the creation cycle which happens on more than one occasion in per year. Be that as it may, the ranchers adapt up to changing situation to acquire advanced change the horticulture by executing AI. It's just a single illustration of AI changing agribusiness, an arising pattern that will help spike a horticultural insurgency. We should be more ingenious this time around.

2.3.2.4 Robots in horticulture

Automated aeronautical vehicles (UAVs) or automated ethereal structures (UAS), in any case called machines, in a mechanical setting are automated airplanes that can be distantly controlled [148]. They work in conjunction with the GPS and others sensors mounted on them. Robots are being executed in agribusiness for crop wellbeing observing, water system gear checking, weed distinguishing proof, crowd and natural life observing, and fiasco the executives [149]. Distant Sensing with the utilization of UAVs for picture catching, handling, and investigation is having a gigantic effect on agribusiness [150]. The rustic business seems to have gotten a handle on meander aimlessly advancement with extraordinary excitement, using these pushed instruments to change current rural strategies [151]. The total addressable assessment of robotization energized plans in each and every important industry is basic – more than USD 127 billion, as demonstrated by a continuous PwC examination. They can be diverged from an ordinary easy to utilize camera for indisputable pictures, yet while a standard camera can give some information about plant improvement, consideration and various things, a multispectral sensor broadens the utility of the methodology and empowers ranchers to see things that can't be found in the recognizable reach, for instance, dampness content in the dirt, plant wellbeing checking. These could help rout the various limitations that hinder agrarian creation. The advancement of the UAS is consolidated with Wireless Sensor Networks (WSN). The information recuperated by the WSN empowers the UAS to propel their usage for example to limit its splashing of engineered mixtures to deliberately doled out areas. Since there are unexpected and ceaseless changes in natural conditions the control circle should very likely react as quick as could sensibly be considered typical. The compromise with WSN can help toward that way [152]. In exactness farming, UAVs are fundamentally pertinent for horticulture activities, for example, soil and field investigation [153], crop observing [154], crop stature assessments [155], pesticide Spraying [156], given in Table. 2.3. However, their equipment executions [157] are absolutely disciple on basic viewpoints like weight, scope of flight, payload, setup and their expenses. An examination including advances, techniques, frameworks and limits of UAVs are analyzed [158].

Table 2.3 Applications of drones in farming

S. No.	Application	Technologies/algorithms used	Results
1	Pesticide Spraying	Wireless Sensor Networks, Gyroscope and Accelerometer sensors	N/A
2	Crop Monitoring, Mapping, and Spraying	DJI Phantom 3 Advanced UAV and other software	UAV could be used to detect faults and identify potential problems.
3	Crop Monitoring	Multispectral sensor	Linear regressions for NDVI with plant nitrogen, aboveground biomass, etc. This may be indicative of good management practices and techniques.
4	Pesticide Spraying	Spray motor	Worked satisfactorily during experiments on peanuts and rice fields.
5	Remote Sensing	Multispectral camera	The UAV remote detection system was tested on a sod field and was capable of doing so.
			Monitoring the temporal changes in the area.
6	Remote Sensing	Spectral Spatial classification, Bayesian information criterion (BIC)	Manual detection of the tomato is difficult, which makes it possible to use this technology; the zones could be classified into tomato regions rather than tomatoes. Detection was carried out successfully on two example images.
7	Crop Monitoring	Hyper spectral Frame	The camera theft campaign was successful in providing hyperspectral information. This allows the concentration of N in rice to be controlled.
8	Crop Monitoring	Camera and Software	Specific manner of monitoring various aspects

	Reinbeck and		of the farm, such as the creation of a digital field map,
			detection of plant health problems, etc.
9	Precision Agriculture Monitoring	–	Provides an approach to segregation of sparse and dense zones in a sugar cane. field. It utilizes satellite data. The precision was 87 percent for the tests.
10	Spraying Fertilizers and Pesticides	Accelerometer and Gyroscope Sensors, Arduino	It possesses the ability to shorten time and human endeavor.

About in excess of 250 models are investigated just as summed up to pick a suitable UAV in farming [159] indicated in Table. 2.4. The farming robot market is relied upon to develop more than 38% in coming years. It is accepted that the requirement for effective horticulture is simply going to turn out to be more significant because of expanding populace levels and changing environment designs [160].

Table 2.4: Arrangement of Drones for Farming Solicitation

S.No	UAV	ROTARY WINGS	FIXED WINGS
1	Flight period	Fly up to 19 min	Fly up to an 60 minutes
2	Wind pressure	Can be operated by wind. 20 to 50 speed gusts	Fly back and forth from the wind for satisfactory descriptions
3	Tractability in mutable course	Allow a new directorate to continue Flight for re-charge.	Allow new directories to be downloaded.
4	Worth range Deployable	$500 to $100,000	$500 to $100,000
5	Assortment	Extremely deployable	Extremely deployable

2.4 Conclusion:

Agriculture plays a significant role in the economic sector. The automation in agriculture is the main concern and the emerging subject across the world. The population is increasing tremendously and with this increase the demand of food and employment is also increasing. The traditional methods which were used by the farmers, were not sufficient enough to fulfill these requirements. Thus, new automated methods were introduced. These new methods satisfied the food requirements and also provided employment opportunities to billions of people. Artificial Intelligence in agriculture has brought an agriculture revolution. This technology has protected the crop yield from various factors like the climate changes, population growth, employment issues and the food security problems. This main concern of this paper is to audit the various applications of Artificial intelligence in agriculture such as for irrigation, weeding, spraying with the help of sensors and other means embedded in robots and drones. These technologies saves the excess use of water, pesticides, herbicides, maintains the fertility of the soil, also helps in the efficient use of man power and elevate the productivity and improve the quality. This chapter surveys the work of many researchers to get a brief overview about the current implementation of automation in agriculture, the weeding systems through the robots and drones. The various soil water sensing methods are discussed along with two automated weeding techniques. The implementation of drones is discussed, discussed in this chapter. The rural business faces different difficulties, for example, absence of compelling water system frameworks, weeds, issues with plant checking because of harvest stature and extraordinary climate conditions. Be that as it may, the exhibition can be expanded with the guide of innovation and subsequently these issues can be settled. It very well may be improved with various AI driven methods like distant sensors for soil dampness content identification and robotized water system with the assistance of GPS. The issue looked by ranchers was that exactness weeding methods conquer the enormous measure of yields being lost during the weeding interaction. Not exclusively do these self-governing robots improve productivity, they additionally decrease the requirement for pointless pesticides and herbicides. Other than this, ranchers can shower pesticides and herbicides adequately in their homesteads with the guide of robots, and plant

40

observing is likewise not, at this point a weight. First off, deficiencies of assets and occupations can be perceived with the guide of man-made mental aptitude in agribusiness issues. In customary methodologies tremendous measure of work was needed for getting crop attributes like plant tallness, soil surface and substance, as such manual testing happened which was monotonous. With the help of different frameworks analyzed, snappy and non-harming high throughput phenol composing would happen with the potential gain of versatile and beneficial movement, on-demand admittance to data and spatial objectives.

Chapter 3
STUDY OF MACHINE LEARNING in IOT
BASED AGRICULTURE

3.1 Introduction

With the presence of development in this high level world, we individuals have extended our limit of the thinking cycle and are endeavoring to join ordinary cerebrum with a phony one. This procedure with examination delivered a very surprising field fake understanding. It is the communication by which a human can make a keen machine. Man-made insight goes under the space zone of programming which can have the alternative to notice its milieu and should thrive to enhance the speed of achievement. Man-made insight should have the alternative to oversee work subject to past learning. Significant learning, CNN, ANN, Machine learning are certain zones which improves the machine work and helps with developing a greater advancement development. The term IOT is explained as "thing to thing" correspondence. The three crucial targets are correspondence, computerization and cost saving in the structure. In [161] gives the start to finish use of IOT in the field of agribusiness and how it might be helpful to individuals. PC based insight has penetrated in clinical science, preparing, account, agribusiness, industry, security, and various regions. Use of AI incorporates learning pattern of machines. This gains us to a sub-space this AI field "simulated intelligence". The sole purpose behind AI is to deal with the machine with data from past experiences and quantifiable data so it can play out its selected task to deal with a particular issue. There are various applications which exist today which fuses analyzing of data from past data and experience, talk and face affirmation, environment assumption, clinical diagnostics. It is an immediate consequence of AI that the region of enormous data and data science has created to an especially unprecedented degree. Man-made intelligence is a mathematical method to manage create watchful machines. As AI strengthened, various new reasoning's and methodology were envisioned and discovered which makes the pattern of basic reasoning simpler. Such strategies are recorded under.

1.   Cushioned reasoning

2.   Counterfeit neural associations (ANN)

3.   Neuro-fleecy reasoning

4.   Expert systems

Among these, the most extensively used and constantly applied system for research expectations is ANN. Our human psyche is the most eccentric piece of the body. Considering the bury associated neural associations, electric signs explores through the neurons with the help of axons. Neural associations which are at the completion of each center point pass the sign ahead. ANN strategy was planned by recalling a comparable thought of the working of the human brain. There are various figuring of this system, for instance, for setting up this particular model computations like Silva and Almeida's count, Delta-bar-delta, Rprop [162], The Dynamic Adaption estimation, Quick prop are used subject to its application. 9 neurons are used at the same time. ANN is an endeavor based methodology which encourages the system to work subject to some inbuilt task instead of a conventional computational adjusted task indicated in Fig.3.1.

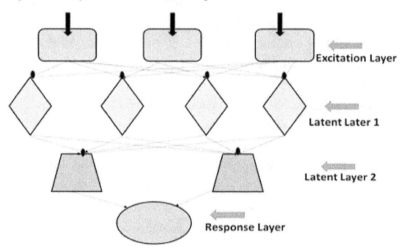

Figure 3.1 AI based methods for Smart farming

The designing of ANN contains three layers:

1.   Data layer
2.   Concealed (focus) layer
3.   Yield layer

Feed forward back expansion part and its limits are showed up above: Input Layer–7, yield layer-1, covered layer-50, number of cycles – 1200. Inception layer-Sigmoidal limit in concealed and yield layer, straight limit in data layer. Moreover, artificial information and AI are by and large hypothesis and theories. These are tweaking and counts. For the execution of these estimations and reasoning based thoughts, there should be a hardware programming interface. The system through which this can be refined is "Embedded structures". Embedded structures are hardware built systems involving memory chips with custom programming changed in it, shown in Fig.3.2.

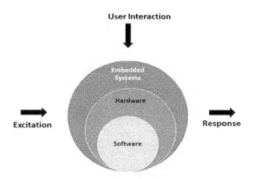

Figure 3.2 Embedded structure for Smart farming

This work consolidates the associations which make introduced systems and AI conscious with the cultivation zone. Execution of AI and expert systems in agribusiness is a scarcely portrayed subject. This topic was analyzed by [163]. Cultivating is the principal piece of any country. At present South Korea, China, North America are placing trillions of money for development in the cultivation region and completing greater improvement progresses. The general population is growing at an astoundingly high rate which is clearly related to the addition in the premium for food. India is a rich focal point for food crops and especially for species. The agriculture region is maybe the most sensitive zones of the Indian economy, supporting any excess zones and spreading its importance in clearing zones. With the happening to advancement in various organizations, it is an amazingly critical feature realize automation in cultivating. The pressure on the agribusiness region will increase with the procedure with improvement of the human people subsequently agri-development and exactness developing have procured a great deal of importance nowadays. This are moreover named as cutting edge cultivation which suggests the usage of hitech PC systems to determine different limits, for instance, weed disclosure, crop assumption, yield recognizable proof, crop quality and much more AI strategies [164]. This paper looks at about the different employments of ANN, ML, and IOT in agriculture and various models which helps in precision developing.

3.2   Writing study

Over the span of ongoing years, there has been an achievable improvement in man-made thinking due to its generosity in the application and is certain in each field. One such field is cultivation. Cultivating appearances various challenges reliably and isn't smooth running business. A bit of the substance issues looked by farmers from seed planting to social occasion of yields are according to the accompanying:

1.   Gather sicknesses attacks
2.   Nonattendance of limit the heads.
3.   Pesticide control
4.   Weed the chiefs
5.   Nonappearance of water framework and leakage workplaces.

Man-made consciousness and Machine learning has entered every single classification referenced previously. In [165], isolated headways in AI class savvy and gave a concise outline on different AI methods. PCs and innovation began infiltrating in this area from 1983 onwards. From that point forward, there have been numerous proposals and proposed frameworks for improvement in farming from the information base to dynamic interaction. Sifting through each cycle, just AI based frameworks have end up being the most possible and dependable one. The AI based technique doesn't sum up the issue and gives a specific answer for a specific characterized complex issue. The writing overview covers significant forward leaps in the area of horticulture from mid 1980s to 2018. The paper examines in excess of fifty headway in advances in the sub space of horticulture. First it examines entrance of artificial neural organizations and master frameworks to tackle previously mentioned issues, at that point AI and fluffy rationale framework. Finally it covers computerization and IOT in the horticulture.

3.3   Artificial neural networks in agriculture

Counterfeit neural organizations have been joined in the agribusiness area commonly because of its focal points over conventional frameworks. The primary advantage of neural organizations is they can foresee and estimate on the base of equal thinking. Rather than altogether programming, neural organizations can be prepared. The creators in [166] utilized ANN to separate weeds from the yields. In [167] utilized neural organizations for guaging water assets factors. In [168] united master frameworks and artificial neural organizations in foreseeing sustenance level in the yield. Customary ES (Expert frameworks) have significant settings when it is being executed. Utilization of ANN makes it up to all glitches of ES. The entire framework is based on a solitary chip PC. Neural organizations consistently end up being the best with regards to foreseeing strategies. Neural organizations can foresee the mind boggling mappings if a dependable arrangement of factors are taken care of. To avoid the issues of ice arrangement in the fields of the island of Sicily, in [169], built up a forecast model utilizing neural organizations. The model is first to take care of with the crude information like stickiness, temperature, precipitation, overcast cover, wind heading (all these information were taken from 1980 to 1983). At that point, the information assembled got changed over into parallel information. These information, presently are partitioned into two strings (info and yield for the neural organization model). The back-proliferation network was utilized as a neural organization indicator. A sum of 10 preliminary sets were created and prepared by the model at first. The ice was anticipated all the more proficiently when a scope of estimations of boundaries (referenced above) was taken as opposed to single

46

qualities. Inside the range of three years, two master frameworks had been created to build the creation of cotton crop. In the first place, COMAX. In 1986, Lemmon made a fruitful endeavor in building up a specialist framework called Comax (Cotton Management master). Lemmon, being the pioneer in AI in horticulture area, built up a program called Gossym which is microcomputer agreeable and support the utilization of Comax [170]. For the first since forever, the master framework (Comax) was effectively coordinated with a PC model (Gossym) and reenacted for the development of cotton crops. This master framework was created to work constantly over time in cotton crop fields. Comax takes three boundaries of the field into thought; planning of water system, keeping up nitrogen content in the field, and development in the cotton crop. Second, COTFLEX. Another master framework for the cotton crop was created by Stone and Toman (1989). The framework was named as COTFLEX. The framework was made dealt with Pyramid 90× PC which utilized UNIX as its working framework. The framework joined the field and ranch data sets to give significant data in regards to the cotton harvest to the rancher so it turns out to be simple for the producer to take basic and strategic choices. The framework created in Texas, and it made recreation models and data sets in the standard based master framework to help Texan ranchers take reasonable financial and remunerating decisions. After powerful testing, COTFLEX was imported to IBM microcomputer and was made open for the use. In [171], the makers inspects about the soyabean crop advancement model which is described as SMARTSOY in the paper and the model is called SOYGRO. The model is explained by data based procedure where it is disconnected into two strategies at first being the positivistic strategy communicating the undertakings to duplicate the patterns of zone experts to arrive at a goal while the ensuing system being the normalizing approach which attempts to duplicate the closures excepting the patterns of the space trained professionals. The damages achieved by dreadful little animals are constrained by orderly technique for choosing the mischief rates and the cost control. Here, the positivistic and regularizing approach breakdowns considering the way that the later method causes it the decision of bug toxic substance and application rate. Regardless, the efficient methodology doesn't help with finding the bug hurt rate on yield. This is a critical disadvantage for making recommendation since we need to mix both the system as the yield decline is controlled by past experiences of the amount dreadful little creature attack, pesticides used and the aftereffects of the yields in the long run. The target to create proposition expressly for soyabean crop relies upon figuring of the mischief rate and the cost to treat the plant and gain the yield. This assessment is gathered by both the techniques. An expert system reliant on cushy reasoning was made [172]. The system was arranged unequivocally for Soybeans crop. This structure gathered its data base from provincial authorities, conveyed composing, and experts of soybean crops. Soft reasoning was considered in looking at the whole structure and inciting the farmer as a trained professional. It was disconnected into five modules. The rule purpose of developing this expert system was to help the farmers in the area increase their soybean creation. The system used MATLAB as a UI module. Examiners developed an expert structure which helped the farmers with when to sprinkle insecticides on the apple normal item to keep an essential separation from the damage in light of bugs and air. The structure was named as POMME. Close by the time, it also admonished the farmers what to sprinkle. As opposed to speculative characteristics from the defilement table, here apple scab disorder cycle model was used in POMME. The delayed consequences of the structure were pleasant and system was certified by the experts who had used it being examined premise [173]. A technique is suggesting the usage of ANN counts for crop assumption in mobile phones had been viably attempted in 2016 by [174]. An assumption model was made. As referred to over, the assumption model of this system had three layers [175]. The adequacy of the model was dependent upon the amount of the covered layers. As an issue of first significance, the ANN model was developed and arranged using various estimations, for instance, Silva and Almeida's counts, Delta-bar-delta, Rprop, and distinctive other to find the best plan. Experimentation strategy was executed to pick the amount of covered layers. There should be a précised way to deal with look at the decision of some covered layers because the assumption structure's precision is dependent on the amount of hid layers. It was found in the investigation that more the amount of covered layers in the ANN model; the more careful were the assumption. Since the inspiration driving the structure was to make it accommodating for the farmers, it is made on APK stage. The source code was written in Eclipse with Java codes in the strike, and the computation was made using Matlab and ANN instrument compartment. The whole report was then removed on the Android stage with the objective that it might be utilized by cells. Other than prescribing the collect to the farmer, the structure similarly has the additional piece of space of empowering the farmer for the manure to be used if the farmer wishes to use his favored yield. In [176], as shown in Fig. 3.2. Evapotranspiration measure is fundamental for keeping up the consistent quality in the hydrologic cycle, sensible water framework method, and water the board. Limits Elevation, Mean each day Temperature, Max. step by step temperature, Min. consistently temperature, Wind Speed, Relatuve Humidity, Sunshine Hours, Daylight hours, Latitude, Condition coefficient. There are more than 20 set up procedures to choose ET which is dependent upon a couple of limits [177].

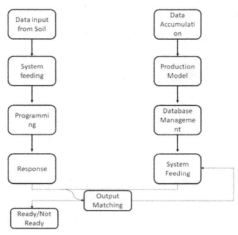

Figure 3.2 Pictorial representation of AI

A critical report was done in the valley of Dehradun; India was reviewing the meaning of the extension of ANN in a couple of strategies for appraisal of ET. Experts amassed month to month climate data from the Forest investigation establishment (FRI) Dehradun for ET appraisal. The methods on which the counts were applied were: 1. Penman-Monteith strategy 2. Levenberg-Marquardt back inducing. It was seen that extending the amount of covered layers in the system achieved instability in the ET evaluation. Along these lines, planning limit with ideal experimentation strategy is to be picked for the as a rule smoothed out appraisal of ET. It was seen that out of six planning figuring of ANN model, work getting ready with 75% data feed in it was precise and had the best number of neurons. In addition, there was a checking between PM strategy and ANN model with the single layer feed forward back spread count. ANN model was arranged and made using Matlab. Six estimations were conjured and assessed. As evapotranspiration is of crucial importance in water framework and water the board, this investigation showed the insightful capacity of ANN structure at whatever point completed precisely. In addition, moreover a method was made to isolate weed from crops with the help of picture assessment and neural associations. It had the accuracy of more than 75% with no ahead of time plant information dealt with into the system. They made expert systems based smart agribusiness structure. The possibility of IoT in this system was to send the data to the specialist with the objective that actuators of the field should have the choice to take reasonable options. For that, the laborer should be sufficiently brilliant to take decisions self-ruling. This system contains temperature, dampness, leaf wetness, and soil sensors. It simply gives the information about the field and doesn't circle back to the water framework measure. In [178] made two ANN models to assess soil suddenness in Paddy fields using very less meteorological data. Both these models were then confirmed and affirmed by pondering saw and surveyed soil sogginess regards. The principle ANN model was created to get the check ET. The help of least, typical, and the most outrageous air temperature was taken. To develop the resulting model, sun based radiation, precipitation, and air temperature data was collected. Both these models achieved the specific and strong appraisal of soil moistness in the paddy fields by using the most un-meteorological data, less work and time use. In [179], the makers look at the neuron stream water framework structures where ANNs were made to anticipate the spatial water dispersal in the subsurface. For spill water framework procedure to properly work, water dissemination in the lower level of the soil is of the grave importance. Here, ANNs makes the assumption which comes accommodating for the customer which in this way achieves the snappy powerful cycle. ANN models give the result of wetting plans (first and second) after the soil is infiltrated with the water from the maker which is outwardly of the land. Subsequently, the ANN model

50

gives reliable guides to the customer. Moreover, researchers developed a model to consider the yield of the maize crop. A multi-layered feed forward ANN (MLFANN) is used. To fuel such sort out, learning computations like GDA (point plunge counts) and CGDA (Conjugate tendency dive figuring) are used. Both the figuring have been formed and impersonated in the MATLAB using neural association toolbox [180]. Precision cultivation and WSN applications join a stimulating new locale of assessment that will unfathomably improve quality in agrarian creation, precision water framework and will have enthusiastic reduction in cost required. Besides, the straightforwardness of association and structure uphold, noticing opens the course for the standard importance of WSN systems in precision cultivating. Using the proposed approach, in finding the ideal sensor topography, we make to cut down execution cost and accordingly make WSN a truly captivating response for a wide scope of fields and advancements.

3.4   Mechanization and remote framework networks in agribusiness

It is basic for any area to develop with time. The agribusiness area needed to adjust the forward leaps and innovations which went along in mechanization field. In [181], it approached with arising research territory of implanted knowledge (EI). Inserted insight in agribusiness area incorporates savvy cultivating, keen harvest the executives, keen water system and brilliant nurseries. It is vital for a country to remember these developing innovations for farming area for development of a country as numerous areas are between subject to horticulture. Additionally, scientists of this paper showed Technology guide (TRM) which thusly explains the doubts with respect to the regions of agribusiness referenced above (savvy cultivating, brilliant water system and so forth) Mulling over the socio and financial imperativeness of horticulture built up a framework which anticipated grape infection previously. Any oddity in the grape plant was seen exclusively after it was tainted and this had an extensive falling apart impact in general grape plantation. The framework utilized different sensors, for example, temperature sensor, leaf wetness sensors, and stickiness sensors in the grape plantation. These sensors send the information detected to the data set in the ZigBee worker which is connected to the sensors. Arrangement of Wireless System Network (WSN) in any field needs to

51

fulfill certain models and Zigbee coalition has created open worldwide principles called ZIGBEE. Zigbee compliances of four layers in particular actual layer, medium access control layer, network layer, application layer as expressed. The three gadgets: Zigbee Co-organizer (ZC), Zigbee Router (ZR), Zigbee End Device (ZED) has diverse capacity in the WSN. The creators in [176] talk about start to finish approach of Zigbee in horticulture. The worker will store the information. The worker is dispatched with a concealed Markov model calculation in it. This calculation is available to prepare the ordinary information detected by the sensors and report any distortion in temperature, mugginess or leaf wetness which can bring about grape illness to rancher by means of SMS. AI is mixed in the framework in advance for adroit derivation of sickness in grapes. The extra preferred position of this framework is it additionally recommends the rancher pesticides and placates manual exertion in the identification of infection. While a comparable strategy for AI was utilized in checking the development of Paddy crops. This framework was created for expanding the yield and efficiency of paddy crops. It likewise end up being practical and strong. In as demonstrated in Fig. 3.3. The sensors utilized in Fig. 5 are for checking farming field are appeared in Fig. 3.4. Sensors, for example, MQ4 and MQ7 are utilized for Natural Gas detecting and Carbon Monoxide detecting individually. DHT11 is utilized for Temperature and Humidity checking of the climate, soil dampness sensor is utilized for estimating soil dampness level and have ceaseless observing.

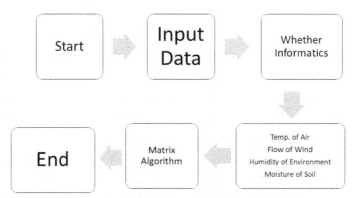

Figure 3.3 Pictorial explanation of evaporation process

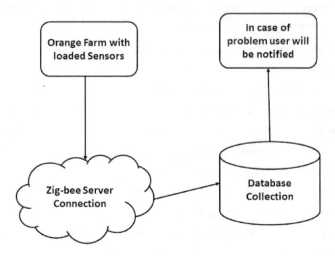

Figure 3.4 Detection of disease using AI

Esp8266 is a wifi module which helps in correspondence between the equipment framework and the gadget which clients use. In one of the examination led in Ankara, Turkey, executing IIS (savvy water system framework), various positive advantages were noticed, for example, less dampness and temperature weight on soil, proficient water utilization, and ignoring human intercession if there should be an occurrence of flood water system. The created framework chips away at three units. Base unit, Valve unit, and Sensor unit. The entire framework is fueled by sun based boards. After the fruitful establishment of each unit, BU will send the location to which the information is to be shipped off SU. Sensors from the SU will detect the dampness content and send the recognized information to a particular location in the BU. Whenever required, BU will impart a sign to VU so it can adjust the situation of the valve to give the dirt water. Notwithstanding, site-explicit utilization of programmed water system framework took birth in the mid-21st century; this technique end up being a critical accomplishment as it diminished the expense, possibility, and intricacy of the created framework. Moreover, the unit can be set up which ships the composts and pesticides in the field utilizing a similar strategy. For that, new sort of sensors would need to align for sending exact data. In [177], research has been led to test the ET based, ICT based, and IIS based innovation. In this work, an exploration was directed in Wheat and Tomato field in which both sprinkler and dribble water system strategy were utilized and tried with ICT just as IIS. A chart of water profundity versus development time of the harvest (week after week) was plotted for every one of the three techniques. A compact perception inferred that IIS was definitely more plausible in a matter of water utilization than ICT and ET based framework. It soar the moderation of water utilization from 18% to an astounding 27%. The utilization of Losant stage for checking the cultivation farmland and individual the farmer by methods for SMS or email if any irregularity is seen by the system. Losant is an essential IoT based most astounding cloud stage. It offers continuous impression of data set aside in it paying little mind to the circumstance of the field. An automated water framework system which uses the GPRS module as a particular contraption. The structure is changed into a chip based section which controls the water sum. It was shown that water speculation reserves were 90% more than the customary water framework structure. AI utilized a passed on far off association for recognizing and control of water framework measure from an inaccessible zone. To improve capability, effectiveness, overall market and to decrease human mediation, time and cost there is a need to divert towards new development named Internet of Things. IoT is the association of devices to move the information without human incorporation. Hence, to secure high benefit, IoT works in agreeable energy with agribusiness to get splendid developing. In The IoT in agribusiness that prompts smart laying out. Use of distant correspondence has changed the rules of correspondence these days and this can in like manner increment the assumptions for cultivating robotization. It has bifurcated the IOT entry into different centers, for instance, actuator, sensor, interface and distant association which offer assistance to correspondence between them. Repeat appraisal and the exchange speed need for this correspondence has moreover been clarified which can be productive for robotization. Execution of WSN (Wireless sensor association) in the agriculture zone and its different systems is portrayed by this paper. Different IEEE rules portray sensor associations, standards like IEEE 802.15.1 PAN/Bluetooth, IEEE 802.15.4 ZigBee and much more are critical to know while orchestrating its application. Investigators similarly inspected about IPV6 the Internet Protocol for far off correspondence and moreover much hardware structure

54

for developing a WSN. By using WSN, Precision developing is possible and the method is used for crop the chiefs. Diverse information is recorded by the sensors and put away in the framework. The framework is made to take in by the past information from the sensors and future moves are made likewise. The creators considered three soil dampness sensors in Pea can edit field and arrived at a resolution that the sensors utilized need site explicit adjustment to bring exact outcomes. The featuring highlights of the paper introduced by incorporates shrewd GPS based distant controlled robot to perform undertakings like; weeding, showering, dampness detecting, fledgling and creature terrifying, keeping cautiousness, and so on Besides, it incorporates brilliant water system with keen control dependent on continuous field information. Thirdly, shrewd distribution center administration which incorporates; temperature upkeep, moistness support and burglary location in the stockroom. Controlling of every one of these activities will be through any far off keen gadget or PC associated with Internet and the tasks will be performed by interfacing sensors, Wi-Fi or ZigBee modules, camera and actuators with miniature regulator and raspberry pi. Warm Imaging is a noncontact and nonintrusive procedure which examination the surface temperature of the horticultural field and gives important criticism to the rancher. In [178], it was talked about the utilization of cloud based warm imaging framework which helps the water system by joining the presentation of the gear's and decide the region of field which requires the water most. The shortfall of consistency will hamper the harvest development and warm imaging can assist with merging this misfortune. Likewise, Thermal Imaging is put to use in agribusiness area really on account of its wide application. The paper by [179] examines different utilization of warm imaging like Pre-reap tasks, Field nursery, Irrigation booking, Yield Forecasting, Greenhouse gases, Termite Attack, [180]discussed the utilization of robot in the farming field. The robot is intended to follow the track of white line where really there is a need to work and other surface is considered as dark or earthy colored. Working of robot is for splashing of pesticide, dropping of seed's, water supply and furrowing. In 2016 a gathering of scientists concocted e-Agriculture Application dependent on the system comprising of KM-Knowledge base and observing modules. The frameworks created in IOT and Cloud Computing underscores on dependable models to give

55

ideal data from the field over 3G or Wi-Fi. TI CC 3200 (RFID) Launchpad was utilized to fabricate the model with other fundamental gadgets. Information base has advantage over customary IOT based frameworks; Knowledge Base is developed to store huge complex organized and unstructured data to help ranchers or even a person with no earlier information on cultivating. In any case, discovering right data in a suitable way is troublesome where giving important information ought to be dispersed in a coordinated and complete way, yet additionally in total way. The information based foundation permits adjusting the adjustments in agribusiness for a superior expansion and adding warning administrations [181]. As level of computerization is needed in every single field so the human intercession turns out to be less and it is vital to plan a format in the beginning phases of the mechanics and hardware. Weed the board is the issue which ranchers face a ton and PC vision can assist with addressing the issue. There is specific contrast between a weed and the ideal harvest. CNN can help to recognize among them and tell us to cut just the superfluous plant. CNN has numerous calculations which can even be utilized to distinguish plants and get the information likewise for ranch [178]. R-CNN widely utilized in article discovery and in mechanization it is utilized for natural product recognition and checking of natural products. In [44], examine the utilization of R-CNN in natural product identification of plantations, while preparing the contribution to the organization is 3 channel shading picture (BGR) of discretionary size. They have utilized VGG16 NET with 13 convolutional network and furthermore ZF network which has 5 convolutional layers. Information increase is utilized in light of the fact that it helps in falsely developing the dataset and changing the changeability of the preparation information. The outcomes talked about by them are promising concerning both mangoes and apples Faster R-CNN outflanked the ZF network approach. Cloud based choices and backing in the agribusiness is blasting now a days. The Decision backing and Automation framework (DSAS) helps the ranchers of the producers to control all the applications through its online interface. DSAS as various stages where the it can interconnect numerous gadgets on the single time and give the constant information to the rancher. The rancher assumes the fundamental part as he can screen the continuous information and furthermore control all the machine through software's. Frameworks

56

like shower regulator will splash the pesticide on the field in a characterized sum. Additionally, water system regulator assists with overseeing water system and compost regulator oversees manure. DSAS works through the information given by various sensors like soil dampness sensor, nitrogen sensor, and so on. In [179], it utilized richness and pH meter to take out the level of elements of the dirt and created remote sensor based trickle water system framework. In [180], it utilized IC 89c52 microcontroller to assemble a shrewd water system framework. The model supplies water just when mugginess and dampness dips under a standard chose esteem consequently it moderates water partially. A self-loader water system framework was created and tried on the field of Okra crops. The framework utilized four dampness sensors and PIC16F877A processor was utilized. The valves in the framework turns ON just when there is a voltage drop across any two sensors in the documented dips under a fixed worth and stays ON until the worth goes to the chose limit esteem [181].

3.5   Implementation of fuzzy logic systems in agriculture

FK-based fuzzy model was implemented to decide the land suitability [178]. Various fuzzy sets were generated using farmer's as well as scientific knowledge congruently. The sets used S-membership functions and were used to determine soil texture, slope, and colour. The research work was done in several villages of Nizamabad district of Andhra Pradesh state of India. In FK-based fuzzy factor maps, it is not necessary to take the lower and upper limits 0 and 1 respectively. This is because in knowledge-driven fuzzy modeling there is no constraint on choosing the membership functions as long as the functions are in context of the factor which has to be modeled (in this case FK-based model). Another implementation of fuzzy modeling was done for land leveling by [179]. They employed fuzzy control theory in the controller of the system. By implementing fuzzy control theory, a precision based result was obtained. High accuracy fuzzy control theory translates the variables (the deviation in the height of the field and expected the height of the field) into the defined variables sets (E and EC)which contains fuzzy terminologies such as 'High', 'Very high' and so on. There are nine sets defined for variable set E and two for variable set EC. This theory helps the controller deduce the position of the bucket which is in turned will be the height of the field. The bucket receives the signal from the receiver. In [180], developed an innovative system for grading the leaf diseases. The system was segregated into five parts namely Image acquisition where the researchers have captured images of Pomegranate leaves, image pre-processing where the captured image is then resized, filtered, and processed according to the required parameter. Then comes colour image segmentation where k-means clustering is used to isolate the healthy part of leave with the disease infected part. Afterward from the resized image, total leaf area is calculated, and with the help of the third part, infected disease area of the leaf is calculated. Finally, in the last part, with the help of a fuzzy inference system, accurate grading of the disease can be taken out. FIS (Fuzzy inference system) was developed by [181] to determine optimum rates of N fertilizer on the basis of field and crop features. Also, [174] implemented FIS to estimate stem water potential. In [175] applied FL model in determining the quality of Apple fruit. In developed fuzzy logic based air controllers to maintain the temperature of storage facilities for Potato. Escobar and Galindo (2004) came up with a simulation software (SCD) which came in handy for many fuzzy based controllers. The software used rule-based knowledge base with IF. THEN condition type. Its graphical characteristics make the software adaptable to any fuzzy algorithm simulations. Another Fuzzy inference system using IF and THEN condition type was developed by [176]. The model forecasted plant disease on the base of weather data. The system was developed to avoid diseases in plant beforehand as disease occurs in specific range of temperature and humidity in the weather. India and China alone constitute 2.7 billion people living under the stress of water shortage. Out of overall water consumption, 70% is consumed in the agricultural process. Remaining is used in infrastructural pipelines and other miscellaneous works. Water leakage is inevitable and uncontrollable in cities. Water demand will shoot up by 50% shortly and this fact cannot be vetoed away. The farming fraternity is the only option in which water usage can be optimized by employing smart irrigation systems. By inculcating smart irrigation system, wastage of water can culminate to a great extent can abridge water consumption by 20%. [177]. There is a major problem of water wastage and a dearth of water in conventional irrigation methods employed. To give an example, Egypt faces a problem of water distribution from the Nile river with neighboring countries. Ample research had been carried out to solve the problems faced in the irrigation, process.

Many companies have developed a sensor-based smart irrigation system. These systems have been developed for optimal water usage, monitoring of water pollution, and to take care of some other grave problems. Soil moisture and temperature sensors interact directly with embedded components in the field and take care of required water distribution among crops without farmer's interaction. Water which is to be fed to the farms, either by the means of smart irrigation or any other conventional method, should be of a good quality. Researchers have started implementing IOT systems and Artificial intelligence techniques in aquaculture sector along with agriculture. The system designed by [178] monitors the quality of water by deploying state-of-the-art automation techniques. In [179], the authors came up with a smart system which controlled valves of sprinklers with the help of temperature and moisture sensors deployed in the field. However, this system did not consider the water pollution problem. In [180], the authors came up with a distributed irrigation system which works on soil water measurement. M2M (machine-to-machine) technology which allows machines to interact with each other autonomously and store the data directly in a cloud-based server online. This M2M technology is in an incipient stage and is developing steadfastly. The authors developed a technology which allows machines to communicate themselves [181]. In [178] the authors presented a complete sensor-based intensive irrigation method which is self-organizing. This system constructed a bottom and upper layer. In [179], the authors tried to demonstrate a prototype of the small-scale smart irrigation system. In [180], the authors considered only automation and IoT in their quest for an intelligent irrigation system. So far, there hasn't been such advent which allows complete freedom of human intervention. This paper tries to bring forward a method through which with the help of AI and embedded technology which eliminates the glitches emphasized in the past.

3.6   Projected Method

Need of automation in the agriculture sector is must and there are many ways it can be implemented in practice. Irrigation is the foremost thing where automation is necessitate for optimal water usage. Soil moisture sensor helps to monitor the moisture level of the soil and starts watering the farm as the value get below the threshold level set by the farmer. The embedded system and Internet of Things help to develop a compact system which monitors the water level of the farm without human interaction. There are many different techniques that we can implement as automation through different forms like using Machine learning, Artificial Intelligence, Deep learning, Neural network, Fuzzy logic. The idea is to use any of these extended methods to reduce human intervention and human efforts. All this methods have their own advantages and disadvantages, but the way they are used differentiate them from each other. The meagre research in the field of deep learning technique which analyses the dataset of images from the past data fed and classifies the plants or flowers. In [181], the authors discusses the Deep learning concepts in the agriculture and the efforts that apply to execute deep learning techniques, in various agricultural sectors. Deep learning application is required in this field as it provides major impact on the modern techniques; it extends the Machine learning by adding more depth into the model. The main feature of the deep learning is the raw data process to increase accuracy and classification. Plant recognition, fruit counting, predicting future crop yield are the main target where deep learning can be implemented. Large dataset of images are required to train the model, while some techniques use text data to train the model. Data source, Data pre-processing, Data Variation and Data augmentation techniques are necessary for the Deep learning to train the model. Future of deep learning in agriculture has many environments and it can prolife rate agriculture sector. In [65] have worked with convolution neural network models and used deep learning in the system by training the model with different images of healthy and diseased plants. Plant disease recognition has a high degree of complexity and so many agronomists fail to diagnose specific disease. The model represented perfectly identifies and gives accuracy upto 99.53%. The idea is to train the model such a way that it identifies plants or flowers when in future any image is fed to the model. To train the model, VGG16 is used as it is the simplest model among all other convolutional networks. This network is characterized by its simplicity, using only 3×3 convolutional layers stacked on top of each other in increasing depth. Reducing volume size is handled by max pooling. Two fully-connected layers, each with 4096 nodes are then followed by a softmax classifier. In VGG16, '16' stands for the number of weight layers in the network. Keras library in python includes VGG, ResNet, Inception, and Xception network architectures. A large image data of different plants and different flowers is used to train the model and check the accuracy. The model then accurately predicts the plant or flower when any random image is fed in the system. This system is necessary in the agricultural sector as every plant has some particular need of environment. A fixed amount of water at regular time and favorable environmental gases around helps the plant to grow perfectly healthy. By classification through deep learning it becomes easy for the farmers or botanist to grow plant, as by identification of plant and its favorable conditions, farmers and botanist can provide such environment and proper irrigation.

3.7 Future scope

The farmers who are young will make more investments in automation with much interest than the elder farmers. The technology which is new has to be introduced slowly with time. Slowly the agriculture sector is moving towards precision farming in which management will we done on the basis of individual plant. Deep learning and other extend methods are used to detect the plant or flower type, this will help farmers to provide favorable environment to the plant for sustainable growth. Eventually the production of more customized fruits and plants will grow, which leads to an increase in the diversity of products and production method. Artificial intelligence techniques are growing at a rapid scale and it can be used to detect disease of plants or any unwanted weed in the farm by using CNN, RNN or any other computational network. Green house farming can provide a particular environment to the plants but it is not possible without human intervention. Here, wireless technology and IOT comes in the run and using the latest communication protocols and sensors we can implement weather monitoring and control without human presence in the farm. Harvesting of fruits and crops can also be incorporated by robots which are specialized in working round the clock for quick harvesting. Application of robotics are vast in farming such as the robots can be used in seeding and planting, fertilizing and irrigation, crop weeding and spraying, harvesting and shepherding. To complete the same work in many cases, it would take approximately 25 to 30 workers. Thermal Imaging can also be implemented by using drones and thermal camera in it. The drones monitor the farm and gives continuous real time data of the field so that the farmers could know in which area of the field the water quantity is less and can only start irrigation in that particular area. This will prevent water flooding or scarcity of water in the field and the crops get advent amount of water all the time. Many different integrated approaches can be used to provide a viable environment and increased growth.

3.8 Conclusion

Agriculture checking is the much require diminishing human interventions basically. Bit by bit revenue for food is showing up at its high apex and the without execution of the bleeding edge procedures in cultivation it is amazingly hard to achieve the extending demand. Agriculture checking is the wonderful concern as it helps with diminishing work and augmentation the creation. Man-made cognizance has been executed in yield decision and to help the farmer in the assurance of the excrements. With the help of the informational collection which the customer has gathered and resolved to the system, the machine talks about among themselves to pick which yield is sensible for harvesting and moreover the manures which advance the most extraordinary turn of events. Significant learning has wide reach and its application in industry has gotten monstrous progress. Using significant learning is an extra piece of room over AI and it adds significance to AI. Various colossal strategies can ensure the farmers with better gathers and genuine field the heads. This in the end helps when all is said in done improvement of the country as food is the central need of any person. IOT indicated its significance to help in the steady checking of the data. IOT is fundamentally used in a savvy watering system. Since, convincing usage of the open new water is essential and with the progress in the development and utilization of computerization water crisis can be tended to. Standard strategies in agriculture have minor effects in this high level world. Water deficiency and flooding both are the difficult issues farmers are standing up to using the customary procedure. Numerous stipulations in this system and the upsetting need to guarantee the cultivating territory prompts the headway of agriculture automation. This paper delivers an arrangement to make a system with the usage of sensors, IOT and AI to automate the standard practices in cultivating. Cultivation robotization is the major concern and emerging subject for every country. The complete people is growing at a very brisk rate and with increase in people the necessity for food augments vigorously. Regular procedures used by farmers aren't adequately sufficient to serve the growing solicitation consequently they need to hamper the earth by using hazardous pesticides in an expanded manner. This impacts the cultivating practice a ton and in the end the land stays pointless with no extravagance. This paper examines assorted robotization practices like IOT, Wireless Communications, Machine learning and Artificial Intelligence, Deep learning. There are a couple of zones which are making the issues cultivating field like collect infections, nonattendance of limit the heads, pesticide control, weed the board, nonappearance of water framework and water the chiefs and this issues can be settled by recently referenced different strategies. Today, there is a sincere need to decipher the issues like usage of hazardous pesticides, controlled water framework, control on tainting and effects of environment in agrarian practice. Motorization of developing practices has shown to construct the expansion from the soil and besides has fortified the earth productivity. This paper surveys created by various researchers to get a succinct layout about the ebb and flow use of automation in cultivating. The paper similarly discusses a proposed system which can be executed in plant farm for bloom and leaf ID and watering using IOT.

Chapter 4

AN INTELLEGENT PROTOTYPE MODEL
FOR SMART AGRICULTURE

4.1   Introduction

Farming is one of the huge wellsprings of the economy in the country. Precision Agriculture is as of now in execution in various countries yet there is a need to realize, improve and advance IoT (Internet of Things) and dispersed processing developments for better production of the gather. There is a steady extension mainstream with people advancement. Modernization in cultivation lessens dependence on individual human work and land. The development grants operational considering and animates dynamic on Farms. IoT licenses us to assemble including data, stock it, devise it, and dissipate the information. The allocation of conveyed registering has experienced a huge climb up the creek without a paddle and would continue filling in the coming future with improved cloud encouraging and taking care of dominances. Artificial Intelligence and IoT is an unfathomable lead as a response for extended benefit. The data through IoT devices is made available publically for research purposes as enlightening records and is taken care of and broke down for extra conjecture related to the yield being made. The Traditional Technique of developing does exclude any cycle, for instance, seed assurance, soil examination, environment assessment, vegetation examination, Nutrient assessment if all of these segments are taken in thought, this all would get an unprecedented change the overall population. The System moreover has a square chain-based conveyance structure to ensure fitting transport with no wastage. Disregarding the way that the pen-paper custom is hard to be displaced at this point restricts a huge load of work, moreover, the assessment of manual work required can in like manner be separated. A more versatile approach to manage IoT things in agribusiness can be tended to by the alleged estate benefit of the chief's structures. They normally consolidate a couple of agribusiness IoT contraptions and sensors, presented in the area similarly as an amazing dashboard with consistent capacities and in-built accounting/uncovering features. Despite the recorded IoT cultivating use cases, some obvious possibilities consolidate vehicle following (or even automation), storing the heads, coordination, etc. With the populace blast, there is a quick expansion in the interest for food and rural stocks and development interaction to improve yield, cost-adequacy, and nature of harvests/rural items being created with new innovation like the Internet of Things (IoT) and Artificial Intelligence. There is a need to build yield, adequacy and upgraded creation of land per unit region taken under thought. It is important to accept new advancements to conquer these issues. There are different advantages related with the institution of new advances which include: expanded profitability, legitimate harvest dissemination, Crop design proposal, appropriate usage of assets, for example, Fertilizers and excrements utilizing the strategy of Automation and AI model. In the advanced summation of the modern upset, 4.0 where we have a restricted measure of assets and their appropriate use is a subject of incredible concern, regardless of whether it's the usage of water or use of minerals from metals this in a roundabout way influences our lives. With the restricted accessibility of assets and expanded utilization there costs have been ascending thus there economical usage is vital. Essentially, on account of Farming where we need to take care of an enormous number of clients, any sort of misfortune at any stage ends up being a colossal misfortune to the economy and the client too. Besides, there is an absence of exploration information in this field. The fundamental rationale is to carry IoT and Machine Applied Farming to India, to plentiful up the specialized utilization of AI and Machine Learning among Farmers, Researchers, and Government. The Motive of the projected proposition is to distinguish and decide the nature and nature of soil situated in a specific region, considering the poisonousness level at present example of time and anticipate its future worth utilizing AI model. The Main goals of the work are:-

i.   We have a restricted measure of assets and their legitimate use is a subject of extraordinary concern, regardless of whether it's the usage of water or use of minerals from metals this in a roundabout way influences our lives.

ii.   Also, there is an absence of examination information in this field, this would create a tremendous assortment of information for the ranchers. Information, huge loads of information, gathered by brilliant agribusiness sensors, for

65

example climate conditions, soil quality, harvest's development progress or dairy cattle's wellbeing. This information can be utilized to follow the condition of your business as a rule just as staff execution, gear effectiveness, and so on

iii.   Better authority over the inward cycles and, therefore, lower creation hazards. The capacity to anticipate the yield of your creation permits you to get ready for better item appropriation.

iv.   Having the option to perceive any irregularities in the harvest development or domesticated animals' wellbeing, you will actually want to relieve the dangers of losing your yield.

v.   Expanded business proficiency through cycle robotization. By utilizing savvy gadgets, you can computerize various cycles across your creation cycle, for example water system, treating, or bother control.

vi.   Improved item quality and volumes. Accomplish better power over the creation interaction and keep up better expectations of harvest quality and development limit through computerization.

4.2   Prototype Model

The arrangement basic format depiction in Fig.4.1 addresses the whole plan of this design. It centers information stream all through the System. The sensor regards used for a comparable cover temperature and wetness of the climate, the moistness content in the soil, its pH state, pungency, NO2 obsession, and a Unified Water System. The data are gotten by Google Firebase where the informational collection resides and the Web Server is encouraged on Apache Web Server.

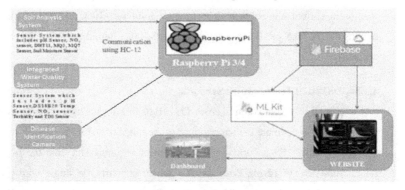

Figure 4.1 Prototype model

66

4.3 Hardware Requirement

4.3.1 Arduino Mega2560 -

The Arduino Mega 2560 is a microcontroller board based on the ATmega 2560. It has 54 digital input/output pins (of which 15 can be used as PWM outputs), 16 analog inputs, 4 UARTs (hardware serial gates), a 16 MHz crystal oscillator, a USB attachment, a power tool, an ICSP header, and a reset key. It comprises everything required to assist the microcontroller; simply connect it to a computer with a USB cable or power it with an AC-to-DC adapter or battery to use.

4.3.2 Temperature and humidity sensor –

DHT22 is a low- cost digital sensor that uses a thermistor to measure the air in the surrounding and it uses a capacitive humidity sensor to measure humidity. They consist of an NTC temperature sensor/Thermistor to measure temperature. A thermistor is a thermal resistor – a resistor that adjusts its resistance with temperature. Technically, all resistors are thermistors – their resistance varies slightly with temperature – but the change is usually very very small and difficult to measure. The humidity sensing element is used, of course, to measure humidity, which has two electrodes with moisture-holding a substrate (usually a pinch of salt or conductive plastic polymer) sandwiched within them. The variation in resistance between the two electrodes is equivalent to the relative humidity. The ratio of moisture in the air to the highest amount of moisture at a particular air temperature is called relative humidity. The typical input voltage range is 3 to 5V and the maximum current allowable is 2.5mA. It is good for -40oC to 15oC temperature readings with +0.5 or -0.5oC accuracy and 0-100% humidity readings with 2-5% accuracy.

4.3.3 DS18B20 Temperature Sensor -

The digital temperature sensor DS18B20 follows a single wire protocol and it can be used to estimate the temperature in the range of -67oF to +257oF or - 55oC to +125oC with +-5% accuracy. The spectrum of acquired data from the 1-wire can range from 9-bit to 12-bit. Because, this sensor supports the single wire protocol, and the regulating of this can be done through an only pin of Micro- controller. This is a superior level protocol, where each sensor can be set with a 64-bit serial code which serves to control copious sensors using a single pin of the microcontroller. It is a part of the integrated Water System. Use a zero before decimal points: "0.25," not ".25." Use "cm3," not "cc."

4.3.4   MiCS-4514 NO2 Sensor –

The MiCS-4514 is a compacted MOS Sensor with two completely self-governing detecting components of one unit. Affectability Factor is resolved as Rs in air separated by Rs at 60 ppm CO. Test conditions are 23oC with a blunder of 5oC and 50 Rh with a mistake of 10%.Sensitivity Factor is characterized as Rs at 0.25 ppm NO2 separated by Rs in air. Test conditions are 23oC with a blunder of 5oC and 50 Rh with a mistake of not exactly of equivalent to 5%.The Silicon Gas sensor structure comprises of an unequivocally miniature machined stomach with an encased warming resistor and sensor board on top. The MiCS-4514incorporates two sensor chips with autonomous warmers and touchy layers. One sensor chip identifies oxidizing gases (OX) and another sensor recognizes decreasing gases (RED).

4.3.5   HC-12 –

HC-12 radio sequential port correspondence module is another age multichannel embedded transmission information transmission module. Its radio working recurrence band is 433.4-473.0MHz, numerous channels can be set, with the venturing of 400 kHz, and there are a sum of 100 channels. The greatest communicating capacity of the module is 100mW (20dBm), the accepting affectability is - 117dBm at a baud stream of 5,000bps noticeable all around, and the transmission distance is 1,000m in open space. There is MCU inside the module, and the client doesn't have to program the module independently, and all straightforward transmission mode is just dependable for getting and sending sequential port information, so it is helpful to utilize. The module uses various sequential port straightforward synchromesh modes, and the client could choose them by AT order as per client necessities. The normal working current of three modes FU1, FU2 and FU3 out of gear state is 80μa, 3.6mA, and 16mA separately, and the greatest working current is 100mA (in communicating state).

68

4.3.6   Raspberry Pi –

The Raspberry Pi is a minimal effort, charge card measured microcomputer that connects to a PC screen or TV, and utilizations a standard console and mouse. It likewise has low force dispersal somewhere in the range of 0.5W and 1W. New Out of Box Software (NOOBS) gives the client a decision of the working framework from the standard conveyances. Raspbian is the affirmed working framework for ordinary use on a Raspberry Pi. In this task, Raspberry Pi basically requests two purposes. To begin with, worker and second for capacity. Apache Web Server is facilitated on this Pi alongside a Firebase data set for the convenience of the information.

4.3.7   pH Sensor –

The pH sensor module comprises of a pH sensor likewise called a pH test and a sign molding board which gives a yield that is proportionate to the pH esteem and can be interfaced straight with any miniature regulator. The pH sensor has an Oxidation-Reduction Potential (ORP) Probe which mirrors the voltage corresponding to the inclination of the explanation to acquire or lose electrons from different substances. This voltage and pH have a lessening reliance; bring down the voltage higher the pH got.

4.3.8   MQ2 Sensor –

Delicate material of the MQ-2 gas sensor is SnO2, which with lower conductivity in clean air. At the point when the objective ignitable gas exists, the sensor's conductivity is higher alongside the gas focus rising. It changes the variety of conductivity over to think about the yield sign of gas focus. MQ-2 gas sensor has a high reasonableness to LPG, Propane, and Hydrogen, additionally could be utilized to Methane and other flammable steam, it is with minimal effort and appropriate for various application. It can gauge the convergence of ignitable gases up to a scope of 300-10000 ppm. It has a customary exemplification of Bakelite.

4.3.9   Turbidity Sensor –

The turbidity sensor identifies water quality by assessing the degrees of turbidity. It embraces light to recognize suspended particles in water by estimating the light conveyance and dispersing rate, which shifts with the measure of complete suspended solids (TSS) in water. As the TTS builds, the fluid turbidity level increments. Turbidity sensors are utilized to direct water quality in waterways and streams, squander water and profluent conclusions, control instrumentation for settling lakes, silt transport exploration and lab estimations. This sensor gives simple and advanced sign yield modes. The edge is versatile when in computerized signal mode. You can choose the mode as per your MCU. On the off chance that you leave the sensor in the unadulterated water, that is NTU < 0.5, it should yield "4.1±0.3V" when the temperature is 10o~50oC. The Operating Temperature of Turbidity meter is 5o~90oC.

4.3.10 TDS Sensor

TDS (Total Dissolved Solids) shows the number of milligrams of solvent solids disintegrated in one liter of water. All in all, the higher the TDS esteem, the more solvent solids broke down in water, and the less spotless the water is. In this manner, the TDS worth can be utilized as one of the references for mirroring the tidiness of the water. TDS pen is generally utilized hardware to quantify TDS esteem. The excitation source is an AC signal, which can viably keep the test from polarization and drag out the existence of the test, in the meantime, increment the solidness of the yield signal. The TDS test is water proof, it very well may be lowered in water for long time assurance. The test cannot be utilized in water over 55 degrees centigrade. TDS Measurement Range: 0 ~ 1000ppm TDS Measurement Accuracy: ± 10% F.S. (25 o C).

4.3.11 Programming Requirement

4.3.11.1   Raspbian OS –

Raspbian is a Debian-based (32-bit) PC working framework for Raspberry Pi. There are various variations of Raspbian including Raspbian Stretch and Raspbian Jessie. In spite of the fact that Raspbian isn't intended to work the Pi like a PC furnishes the clients with a LXDE work area foundation. The Pi ingests not have a lot of processor speed or memory, yet it has adequate help to run LXDE and a portion of the applications like the oversimplified Epiphany internet browser and a few all the more further.

4.3.11.2 Arduino Software –

The Arduino Integrated Development Environment (IDE) is a cross-stage application (for Windows, macOS, Linux) that is written in capacities from C and C++. Arduino is an open-source gadgets stage dependent on simple to-utilize equipment and programming. It is utilized to compose and transfer projects to Arduino viable sheets, yet in addition, with the assistance of outsider focuses, other merchant advancement sheets. The Arduino IDE underpins the dialects C and C++ utilizing uncommon principles of code organizing. The Arduino IDE utilizes the program avrdude to change over the executable code into a book document in the hexadecimal encoding that is stacked into the Arduino board by a loader program in the board's firmware. By getting contributions from sensors, Arduino faculties the climate and influences or impacts its environmental factors with various activities and actuators. Arduino is obliged to assemble and transfer the essential codes to run the UNO.

4.3.11.3 Firebase –

The Firebase real-time Database is cloud-facilitated. Information is put away as JSON and synchronized continuously to each associated customer. At the point when you create cross-stage applications with our iOS, Android, and JavaScript SDKs, the entirety of your customers share one real-time Database event and naturally gather refreshes with the most current information. Firebase real-time Database permits you to create rich, cooperative applications by empowering secure admittance to the data set straightforwardly from customer side code. Information is persisted locally, and even while disconnected, continuous occasions keep on terminating, giving the end-client a responsive encounter. At the point when the gadget recovers association, the real-time Database synchronizes the neighborhood information changes with the distant updates that happened while the customer was disconnected, combining any contentions consequently.

4.4 Results

4.4.1 Proto-type Model 1

The projected model in programming investigation is demonstrated in Fig. 4.2. The offered practice is executing the possibility of IoT in the developing field by using insightful sensors to make agribusiness field a more shrewd one. The fundamental job of the errand is to total data of various center points and to deal with this information. The producers will actually want to control the exercises indirectly through a versatile application similarly as access the records through a cloud. The arrangement of the arrangement is to make unite checking and control for the country land. This can be kept up and worked from any spot distantly using a cell. The application customer can deal with focal organizations of collection of normal, soil, readiness, and water framework data; thusly accomplice such data and channel - out silly data from the chance of evaluating crop execution; and figure crop guesses and modified thing proposition for a particular farm using the application. The structure can arrange basically any IoT device, including financially open sensors, cameras, environment stations, etc and can shape a great deal of these materials which would execute it flexible for a single customer to cover a tremendous region and save their data in the cloud for execution comprehension and recommendations. End-customer can get all of these pieces of the field on Smartphone by an association Smart Agro Services and can coordinate the undertakings. The sensor network is figured to get information concerning the climatic essentials of the property like Soil Moisture, Temperature, Light, and Humidity. With the help of this, the arrangement will choose the system on the field. A singular residence can have various yields divided into fields. So a particular yield will have a couple of limits to be obliged. This, we need to have a pack that will assemble data freely. These centers are presented on various parts on the field dependent upon the limits. Each the center includes a chip Raspberry Pi and a sensor related with it. Sensors may be temperature and sogginess sensor or soil moistness sensors. The soil soddenness sensor being a straightforward sensor requires an ADC (Analog to Digital Converter). The data from the sensor is in the basic design and ought to be changed over to cutting edge structure. In this way the rough data is given to ADC which hence to automated. The high level data is as voltage regard and depending upon voltage regard the degree of clamminess in the earth is taken. The sensors are related with Raspberry Pi. Raspberry Pi accumulates the data from the sensors of that center. There are different center points that are set around the field. Using the gathering advancement the farmer can make definite decisions like in which piece of the field soil moistness has decreased and where to divert the water framework system. Similarly as when to kill on and the motor siphons and various devices for the limits to be kept up. Data from all of these centers are assembled and moved to a cloud. Here, we are using the cloud organization as a limit informational collection. The Data sent off the cloud is taken care of in the cloud data base. Farmers can sign in to their different records to see their arrangement of encounters and the current data of each center. The data from the cloud is given to the adaptable application. With the help of the flexible application, the farmers acquire effortlessness to control various contraptions and record the readings from the sensors. The mugginess and residue focuses given in Fig. 4.3 and Fig. 4.4.

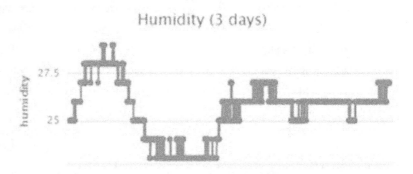

Figure 4.3 Humidity graph

Figure 4.4: Dust Concentration analysis

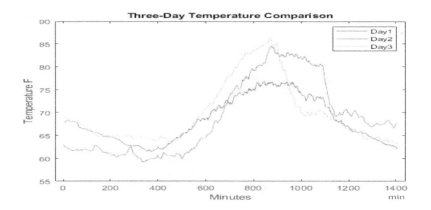

In Fig. 4.5, the mean characteristics and 95% sureness ranges (CIs) over the five cross-approvals overlay were .82 (.81–.83), .78 (.77–.79), .74 (.72–.77), and .76 (.74–.79) for a fixed opportunity to the start, sliding window, sliding window with dynamic thought, and on clinical premium, independently. The AUPRC with mean characteristics and 95% CIs were .385 (.358–.408), .007 (.006–.009), .009 (.008–.010), .014 (.011–.017) for a fixed opportunity to the start, sliding window, sliding window with dynamic consolidation, and on clinical premium, independently.

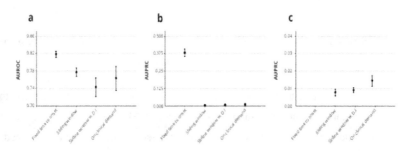

Figure 4.5 Performance analysis

In Fig. 4.6, the ten most significant clinical boundaries for each model of the four outlining approaches are appeared. The boundaries are arranged by the diminishing methods for the supreme SHAP esteems for all people in the dataset. The blue even bars in the left section of Figure 4 show the methods for the total SHAP values. In the nearby clarification synopsis in the right-hand section of Figure 4.6, the appropriations of the SHAP values for each clinical boundary are appeared and shading coded by the boundary esteem related with the nearby clarification.

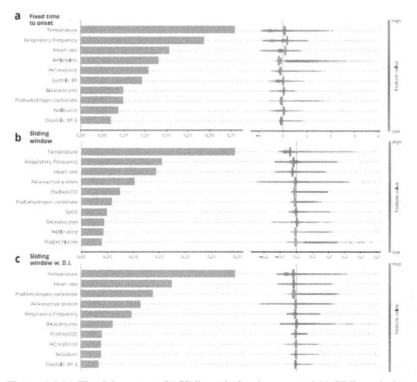

Figure 4.6 (a) Fixed time onset (b) Sliding window is appeared (c) Sliding window with dynamic consideration

This introduced work proposed an IoT and AI-based arrangement for the Agricultural Sector containing an assortment of Micro-controllers, sensors, and a Consolidated Water Quality System. In this technique, the data converged from Sensors will be moved to Cloud for getting ready and Data set relationship through the Internet. The continuous data will be dealt with into a ML figuring after a particular fixed period of time, considering which it would predict the soil condition of the structure. For envisioning the variable limit of Soil, the Regression method would be used at this point checking the Toxicity of Water Anomaly acknowledgment technique is most suitable. This system will be help given to the farmers to digitalizing cultivation. Approximating possible work on this structure may include storing up the data and conveying organizations for each yield exclusively.

4.4.2 Prototype Model 2

The planters are the end-users of this application. This application will provide full radio connectivity to the farmers for their farm. The concocted data from the cloud will be obtained by the farmers using the mobile application. The farmers will get a graphical illustration of data as well for more suitable comprehension of the parameters. With the help of this data summary, the farmers will be aware of the climatic circumstances of the farm and consequently will constrain the accessories such as light and motor pumps. Each farmer will have its own account by which he can log in utilizing an individual username and password. New users can record using their email ID and create a new account. The farmers will get the direct readings from the sensors for the node they have selected. Based on this they will also get an alert for which device should be switched on or off. The operators can thus check for the readings and wirelessly regulate the devices for the field. Hardware results showing monitoring and controlling of moisture level and animal detection is shown in Fig.4.7.The soil moisture sensor senses and measures the moisture level in the soil. The PIR sensor detects the animals and a high frequency sound signal is provided. The ph sensor and water flow sensor is used to optimize the fertilizer usage. These data are processed and the optimum water level will be supplied to the field by automatically switching on the power supply to the water pump. These data will be transmitted to the user's mobile phone through Iot using a separate IP address for the given microcontroller which is programmed to send the data given by the sensor to the user through a web page showing the live condition of the field.

Figure 4.7 Prototype model 2

The reenactment work on the endeavor has been done in PROTEUS PROFESSIONAL programming. The going with picture in Fig. 4.8 shows the noticing and the showing collaboration of various modules application in farming.

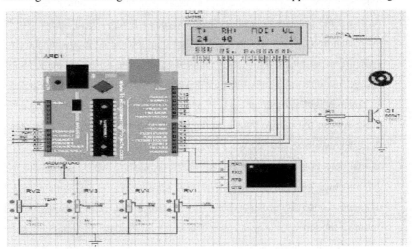

Figure 4.8 Schematic Diagram

Farming is the essential wellspring of food supply for all nations on the planet. Water is the Essential assets for horticulture. The mechanized water system and harvest field observing framework is utilized to enhance the utilization of water asset for horticulture. The framework comprises of sensor network for stickiness, temperature, soil dampness, shading and water level sensors. soil dampness, temperature, water level, shading sensor are put in the root zone of the harvests. The microcontroller of the regulator unit is customized with limit estimations of the temperature and dampness content. The regulator unit is utilized to control the water system engine accordingly controlling the water stream to the field. Notwithstanding that water level sensor is set in this field, on the off chance that it is abundance water the engine gets naturally siphons the water into the external zone. Shading sensor give the suitable shade of leaf and the client give the pesticide prior to annihilating plants. Field measure information about paddy plants. Raspberry pi is utilized in the regulator mode. Web of the things (IOT) is a biological system of associated actual articles that are open through the web. Ongoing checking information can be used and the presentation can be followed. Henceforth high return can be accomplished. This undertaking is essentially centered on improving the farming fields yield by giving a checking framework successful and productive use of water asset. Subsequently further advancement in this venture will prompt a more prominent productivity in the field of farming. The projected hardware implementation is given in Fig. 4.9.

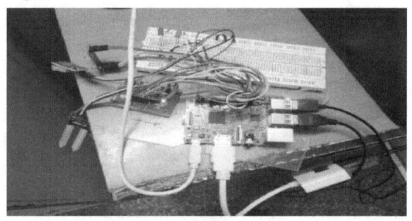

Figure 4.9 Hardware module

Fig. 4.10 shows the monitoring condition simulated in Proteus Professional software version 8. The virtual terminal represents the functional status of the sensors connectivity.

Virtual Terminal

```
HUMAN DETECTED...
HUMIDITY LOW
HUMAN DETECTED...
HUMIDITY LOW
HUMAN DETECTED...
TEMPERATURE HIGH
HUMIDITY LOW
HUMAN DETECTED...
TEMPERATURE HIGH
HUMIDITY LOW
HUMAN DETECTED...
TEMPERATURE HIGH
HUMIDITY LOW
HUMAN DETECTED...
TEMPERATURE HIGH
HUMAN DETECTED...
TEMPERATURE HIGH
```

Figure 4.10 Simulation of output 3

The temperature estimation examination of is demonstrated in Fig. 4.11. High temperature, regardless, for a brief period, impacts crop advancement especially in quiet yields like wheat. High air temperature diminishes the improvement of shoots and consequently decreases root advancement. High soil temperature is more fundamental as damage to the roots is not kidding achieving critical diminishing in shoot advancement.

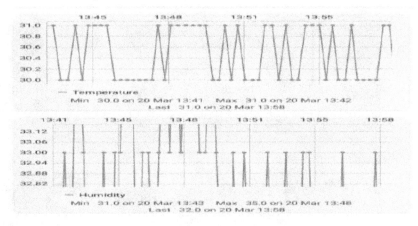

Figure 4.11 Temperature measurement

The humidity measurement analysis of is indicated in Fig. 4.12a and Fig. 4.12b. Plant needs four things to endure: light, water, soil and air. Nonetheless, to raise sound plants, the main component is the impact of water. Relative moistness is a proportion of how much water the air can hold at some random temperature.

79

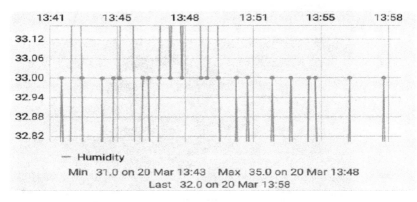

Humidity
Min 31.0 on 20 Mar 13:43   Max 35.0 on 20 Mar 13:48
Last 32.0 on 20 Mar 13:58

Humidity

Figure 4.12 (a) Humidity measurement graph (b) Showing peak

The software output of the projected work is shown in Fig. 4.13. It simply shows the analysis of the of crop for different humidity. It indicates the level of humidity good for crops and also inform about the leveling of watering required for the crops.

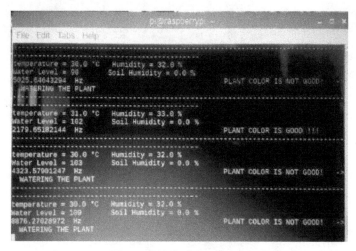

Figure 4.13 Software output

After examining the survey papers on intelligent farming such as IoT based monitoring system in smart agriculture, Smart Farming System using sensors for agricultural task automation, Sensor data collection and irrigation control on vegetable crop using smart phone and wireless sensor networks for smart farm and Remote agriculture automation using wireless link and IoT gateway infrastructure, a novel agricultural automation system using Internet of Things (IoT) is proposed. This system provides real time information about the farmland and alerts the farmer in case of animal threats. The proposed system also prevents the trees from illegal cut down. The future enhancements are given below:

- Irrigation system can be monitored.
- Damage caused by predators is reduced.
- Increased productivity.
- Water conservation.

4.4.3 Prototype Model 3

Identification of plants disease

Plant diseases are considered as one of the most important concerns in the farming sector. Crop diseases can decrease the crop yield. Still, there is no full-proof process to detect the plant disease by a farmer. The detection of disease through

81

optical surveillance is considered as efficient. Hence, it is important to explore novel techniques which can detect and predict the platypus disease, which can also improve the growth rate of crops in the farming sector. Machine Learning (ML) based on computer vision is considered as one of the best choices to detect crop disease. In this work, an ML and mobile grounded framework are proposed to computerise the crop disease (plant leave). In this work, we utilised a Convolutional Neural Network (CNN) for categorising five disease groups. In the first part of the work, a dataset of 295 images of leaves is collected. The condition of the leaves was analysed using the CNN model. The CNN model helps to identify healthy and diseased leaves. The CNN model was integrated with the Android mobile app, which helps to capture images of the infested leaves. The proposed system shows the disease type with the sureness percentage. Hence, the proposed system can detect the disease and help farmers increase their crop productivity. The sample of healthy and diseased plants is taken from our dataset illustrated in fig.1.

Fig 4.14 Images of leaves

The system model is given in fig. 4.15.

- In the first step, we develop a ML-powered framework which is systematised into two portions, implementing it on mobile user devices in the agricultural field and cloud.

- In the second stage, we are capturing, processing, and visualising large imagery datasets.

- Finally, a user-friendly interface has been developed on top of the CNN model to permit farmers to interact with the disease detector conveniently on the mobile side.

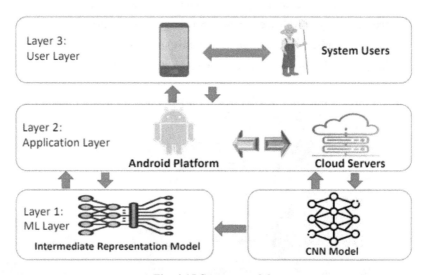

Fig. 4.15 System model

Layer 1 defines the deep learning framework utilised in the CNN model and the Intermediate Representation (IR) framework runs on the mobile device.

Layer 2 demonstrates the user interface, which is established as an Android app to allow users (as indicated in layer 3), to interrelate with the arrangement suitably.

```
Layer (type)                    Output Shape              Param #
================================================================
rescaling_1 (Rescaling)         (None, 180, 180, 3)       0

conv2d (Conv2D)                 (None, 180, 180, 16)      448

max_pooling2d (MaxPooling2D)    (None, 90, 90, 16)        0

conv2d_1 (Conv2D)               (None, 90, 90, 32)        4640

max_pooling2d_1 (MaxPooling2    (None, 45, 45, 32)        0

conv2d_2 (Conv2D)               (None, 45, 45, 64)        18496

max_pooling2d_2 (MaxPooling2    (None, 22, 22, 64)        0

flatten (Flatten)               (None, 30976)             0

dense (Dense)                   (None, 128)               3965056

dense_1 (Dense)                 (None, 10)                1290
================================================================
Total params: 3,989,930
Trainable params: 3,989,930
Non-trainable params: 0
```

Fig 4.16 CNN Structure

4.4.3.2 Data set

We have divided the dataset into three categories such as: training, validation and testing. Table 1 shows the 5 disease classes in 5 crop species. The quantity of images is analysed in the CNN framework. In specific, we gradually tested random mixtures of values till we accomplished suitable outcomes. To increase the training accuracy and minimise training loss of the CNN system, we carried out a sequence of image preprocessing alterations to the training dataset. We also analysed the different features of images, such as colors, added noise, and utilised image desaturation, which makes pixel colours muted by mixing black and white colors. The main objective of the transformations is to deteriorate the impact of the contextual issue throughout the training process. The normalisation of the image was carried out using the following equation:

$$O(i,j) = \frac{I(i,j) - \mu}{\sigma}$$

Where μ is the mean value from each pixel's value I(i, j), and σ is the standard deviation of the input image. Where I and O are the input and output images.

84

Table 4.1 Datasets of Images used in analysis

S.No	Classes of Plant Disease	Training	Validation	Testing	Total
1	Apple scab	50	18	9	77
2	Apple Black rot	32	14	6	52
3	Apple Cedar apple rust	30	19	8	57
4	Apple healthy	40	15	6	61
5	Blueberry healthy	30	12	6	48
	Total	182	78	36	295

To avoid the over fitting issue of the CNN model, we carried out the geometric transformations process to eradicate the positional biases present in the training data.

4.4.3.3. Employment of CNN

The CNN model is employed by utilising Keras development environment 2.4. It is an open-source neural network library. To create the image, we used the keras preprocessing.use the ImageDataGenerator library to augment some of the images in our dataset. The size of the training images must be the same before applying them as input to the CNN. A 4.50 GHz Intel Core TM i7-16MB CPU processor powers the system. The training errors and loss are given by:

$$M = \frac{1}{n}\sum_{i=1}^{n}(y_i - x_i)^2 \quad (1)$$

where M represent mean square error, y determines the predicted class , and x represents the actual class. M determines the error in object detection

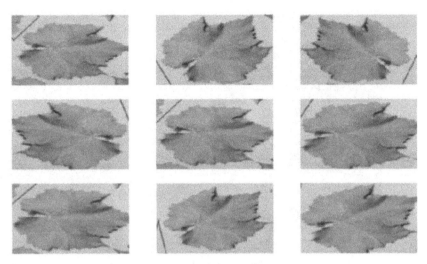

Fig 4.17 Dataset Intensification

4.4.3.4 Mobile App

The plant disease detector's user interface is implemented as a self-contained mobile app developed utilising the Kotlin platform. It allowed us to write a single codebase for the scheme's business logic, and then organize it as an Android app. The pictures of the mobile app for identifying plant leaf diseases are given in Figure 4.18. Figure 4.18a gives a picture screen, which permits farmers to either capture an image of the diseased plant or upload the image (see Figure 4.18b). After a picture is taken, the image is uploaded onto a cloud server to identify the disease classes by using the proposed CNN system. The captured image is transferred to the cloud side via a REST (Representational State Transfer) service in the form of a JSON (JavaScript Object Notation) image object. Figure 4.18b gives an instance of the implication outcome of the CNN system on the mobile app. CNN gives a sureness score of 0.97. The whole operation took around 0.88 seconds. Hence, it can be concluded that the proposed model can be utilised to identify the plant disease.

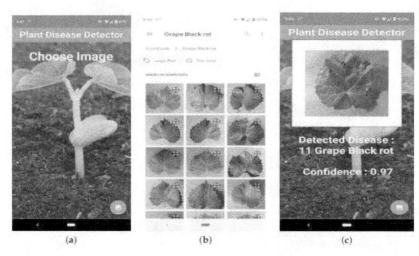

Fig. 4.18 Pictures of Installed App. (a) Landscape Screen Image, (b) Image Collection, (c) Outcome of the proposed system

4.4.3.5 Experiment Evaluation

We have analysed the efficiency of the system in terms of accuracy and performance. To estimate the efficiency of the projected system, we have integrated the model in the Android mobile app and total time to execute the operation is estimated to perform the various operations such as: Image capturing, image transformation, and disease identification processes. For all analyses, several iterations were performed and the average of the iterations was taken. For classification correctness, it is perceived that the projected model gives better outcomes in normal environments although the plant images are taken at a different range from the camera, locations, and brightness circumstances. It is observed in Figure 4.19 that the projected model successfully identifies the fluctuating plant leaf diseases. Figure 4.19a to Figure 4.19d indicate the performance of our projected model, which obtained a high classification in our testing dataset. It is also seen that for some crops, such as tomato diseases, the projected system failed to obtain high confidence levels.

Figure 4.19e gives an instance of a 70% sureness value for identifying the tomato disease. The confidence level is low in this case for several reasons, such as an inadequate training set or due to the presence of noise.

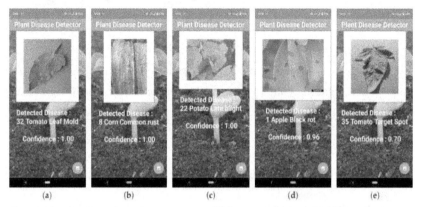

(a)   (b)   (c)   (d)   (e)

Fig. 4.19. Identification of Diseases: (a) Tomato Leaf Mold: (b) Corn Common rust: (c) Potato Late Blight: (d) Apple Black Rot: (e) Tomato Target Spot.

Chapter 5

OUTCOMES & LIMITATIONS

5.1 Objectives of the Study

The objectives of the projected work are as follows:

- Employment of artificial intelligence in agriculture for optimization of irrigation, application of pesticides and herbicides, identifying crop diseases.
- A comprehensive evaluation on computerization in agriculture using artificial intelligence.
- Comprehensive study of Machine Learning applications in IoT based agriculture and Smart farming.

5.2 Outcomes of the Study

- As in different ventures, the use of the Internet of Things in farming guarantees already inaccessible proficiency, decrease of assets and cost, Robotization and information driven cycles. In horticulture, in any case, these advantages don't go about as enhancements, yet rather the answers for the entire business facing a scope of risky issues.
- The present farming is in a race. Ranchers need to develop more items in decaying soil, declining land accessibility and expanding climate vacillation. IoT-empowered agribusiness permits ranchers to screen their item and conditions continuously. They get bits of knowledge quick, can foresee issues before they occur and settle on educated choices on the best way to maintain a strategic distance from them. Also, IoT arrangements in farming present mechanization, for instance, request based water system, preparing and robot collecting.
- A major strength of this study is that we created a language for discussing the concept of framing as essential to building machine learning risk-prediction models. On a large, population-based, open cohort we showed that the same machine learning architecture, applied to the same data, gave rise to many different models when framing is varied. We believe inclusion of this aspect

89

in future risk-prediction models is fundamental to enabling healthcare professionals to discuss the clinical value of developed models or models underdevelopment with technical professionals who construct the machine learning models.

- The study is not without limitations. It is a weakness that we did not test multiple models, such as sequential models, as they are widely applied in sepsis prediction literature. However, it is imperative to stress that even if we had included sequential models, they would still need to be framed and evaluated at specific times, exactly as it was done for the chosen models in this study. Also, XGBoost does not support sequential modelling, thus it would be necessary to introduce a second risk-prediction model with manual imputation of missing values.

- We defined "suspected infection" in accordance with previous studies in the field, but the chosen approach may have led to the misclassification of some patients into the wrong sepsis group (positive/negative). In terms of the SHAP analysis, the addition of sequential models would lead to explanations over multiple time steps that would have to be equated.

- It is observed that how AI hazard forecast model, when applied to a dataset got from general clinical and careful divisions can be outlined in an unexpected way, and how this outlining has direct ramifications for both the model's prescient execution and the model's learning. The significant results of outlining command consideration from clinicians and AI designers, as understanding and detailing of outlining are vital for the fruitful turn of events and clinical execution of future AI innovation. Model outlining should mirror the normal clinical climate. The significance of appropriate issue outlining is in no way, shape or form selective to sepsis expectation, but instead applies to most clinical danger expectation models.

5.3 Limitation of the Study

- The limitation is that this study only concerned sepsis risk prediction in secondary care, and findings may vary in different settings and for other outcomes. However, the fundamentals of how framing is directly linked to the clinical problem, which the machine learning model must solve, do not change.

- Another limitation is that the smart cultivating ceaselessly requires web availability. The non-industrial nations 'provincial part didn't follow those measures and the web is slower. The IoT related hardware permits the rancher to comprehend the utilization of innovation and to learn.

- Security:
i. As the IoT frameworks are interconnected and impart over organizations. The framework offers little control regardless of any safety efforts, and it very well may be lead the different sorts of organization assaults.
ii. Security: Even without the dynamic cooperation on the client, the IoT framework gives significant individual information in most extreme detail.
iii. Intricacy: The planning, creating, and keeping up and empowering the huge innovation to IoT framework is very confounded.
iv. In identification of classes of crop diseases, it is noted that the projected system fails to give a high level of accuracy between the crops having identical leaf phenology. This case may occur between tomato and potato diseases because their leaves have some common characteristics such as color, size, and canopy organization.

CPSIA information can be obtained
at www.ICGtesting.com
Printed in the USA
LVHW031202310123
738241LV00013B/1509